Diagnosis of
Mineral Disorders in Plants

Volume 3

Glasshouse Crops

By Geoffrey Winsor and Peter Adams

new photographs by

Peter Fiske and Andrew Smith

Glasshouse Crops Research Institute
Littlehampton, W. Sussex, UK

General Editor: J. B. D. Robinson

*formerly Long Ashton Research Station, University of
Bristol, UK*

LONDON:
HER MAJESTY'S STATIONERY OFFICE

ISBN 0 11 242723 5

Contents

Acknowledgements

Five of the colour photographs (Plates 14, 72, 74, 77 and 119) were kindly provided by Dr E J Hewitt and Mr E Watson of Long Ashton Research Station.

Most of the remaining photographs were taken at the Glasshouse Crops Research Institute by Messrs P Fiske and A Smith of the Photography and Visual Aids Section, though a few earlier pictures by their predecessors, Mr J Baxendale and the late Mr J E Morton, have also been included. Plates 171 and 181 are from nutritional studies by Mr C Bunt, whilst Plates 43, 99, 101, 128, 136 and 160 are from experiments by Mr D M Massey. Plate 40 is from a photograph made available to us by Dr B Thomas of the Virology Section. The 27 graphs were prepared for publication by Mr M Bone, also of the Photography and Visual Aids Section at GCRI. We are particularly grateful to Miss G A Linfield for her meticulous typing of the text.

Analytical data for molybdenum and sulphur, included in the captions to the colour plates, were kindly provided by Mr R Charlesworth of the ADAS South East Regional Sub-centre, Wye, Kent, as also were some of the data for calcium and magnesium. Further analytical data for calcium, magnesium and micronutrients were provided by Mr M H Adatia.

Finally, may we acknowledge the co-operation of the many members of the Chemistry (later Plant Nutrition and Analytical Chemistry) Department who assisted in the long-term nutritional studies at GCRI from 1957 onwards, and also of the Nursery supervisors and staff responsible for growing the experimental crops; the illustrations and accompanying descriptions in this volume are largely based on their work.

Copyright

CHAPTER 1
Introduction

The principles of plant nutrition outlined in Volume 1 are of wide general application. Although the higher plants may vary somewhat in the details of their metabolism, as for example in the levels of activity of particular enzymes or the accumulation of certain inorganic and organic constituents, their basic nutrient requirements are broadly similar. Thus the higher plants all need a range of nutrients for growth and reproduction, including the macronutrients nitrogen, phosphorus, sulphur, potassium, calcium and magnesium, together with the micronutrients iron, manganese, copper, zinc, boron, molybdenum and chlorine.

On this basis it might seem that the problems involved in meeting the nutritional requirements of the various crops were generally similar, and that to discuss them in a series of separate volumes was merely a matter of convenience. In practice, however, the techniques of growing the main groups of agricultural and horticultural crops vary considerably, and nowhere are such differences more obvious than for 'protected crops' i.e. those grown in greenhouses clad with glass or plastics.

To appreciate the special needs of glasshouse crops, reference must first be made to the environment in which such crops are grown, the levels of productivity achieved, the way in which nutrients are supplied and the rates and timing of application of fertilisers. The picture that emerges is one of considerable specialisation.

The greenhouse environment

The purpose of a greenhouse is to control the environment to provide conditions best suited to the growth and quality of the crop. Whereas in the past the aim was generally to achieve maximum yields, the rising costs of labour and energy for heating or cooling now make it necessary to think in terms of the balance of crop value and production costs i.e. maximum economic yield.

The main features of the greenhouse environment, and the contrast with conditions in the field, with particular reference to temperate zones, are outlined briefly in the sections which follow (see also Winsor, 1968).

Water supply and rainfall. In protected cropping the whole of the water requirement has, of necessity, to be provided by irrigation using water from the mains, natural rainfall catchment or from a borehole. Whilst this inevitably increases costs and may sometimes create problems where the water supply contains excessive amounts of soluble salts or specific components such as sodium or boron, ability to control the water supply to the plants is a fundamental component of intensive crop production. Outdoor crops on the other hand may suffer from either shortage or excess of water, the main problem being that the overall pattern of rainfall is unlikely to match crop requirements throughout the whole period of development. For example, once full ground cover has been established, the water requirement of a crop is largely determined by the incident solar radiation but the distribution of rainfall throughout the year is unlikely to follow any such pattern; for the contrast between water requirement in a U.K. glasshouse and the average rainfall see Winsor (1968: p. 11).

In modern commercial practice the water requirement of greenhouse crops is usually controlled in relation to radiation figures for the preceding day or week, adjusted according to stage of growth, though soil tensiometers have been used sometimes for this purpose.

In addition to ensuring that crop growth is never limited by lack of moisture, except in certain special circumstances where temporary restriction of growth is desirable, the existence of an irrigation system makes it possible to provide nutrients to the crop at any stage of growth. Much of the specialisation involved in the practical nutrition of greenhouse crops is indeed concerned with the formulation and use of nutrient solutions, often referred to as 'liquid feeding'.

Temperature. Protected cropping makes it possible to control air temperatures to favour plant growth. Modern greenhouses equipped with efficient heating systems, large automatic venting systems and even computerised controllers can maintain the temperature regimes best suited to a crop at various defined stages of growth. Furthermore, the growing season, usually limited to only part of the year outdoors, can be greatly extended in a greenhouse, and some crops, notably chrysanthemums, are grown commercially on a year-round basis.

Since nutrient uptake is temperature-dependent it is sometimes advantageous to control root temperature also. Whilst under-soil heating is generally uneconomic due to heat losses downwards, heated benches separated from the underlying soil can be used very effectively for propagation. Hydroponic systems also, being insulated from or raised above the soil, are normally equipped to maintain adequate root temperatures. The uptake of phosphorus is particularly temperature-dependent and deficiencies of calcium, manganese and iron have also been observed in tomato plants grown in nutrient film culture where the solution was maintained at an undesirably low temperature (13–14°C; Moorby et al., 1977). Inhibition of the growth of tomatoes at low root temperatures, and the accompanying symptoms of iron deficiency, could rapidly be relieved by increasing the root temperature without any change in the composition of the nutrient solution (Moorby et al., 1979).

Light. One major consequence of protected cropping is an inevitable reduction in light reaching the crop. Thus, although about 91% of the light reaching clean glass of horticultural quality perpendicular to its surface is transmitted, this value falls as the angle of incidence increases (Lawrence, 1963). Overall light transmission will also be reduced by the structural components of the greenhouse and, although modern designs minimise this loss, the light levels inside will generally be only some 55%–65% of those outside during winter and summer respectively.

During very sunny weather this reduction of light due to the layer of glass or plastic will be of little consequence; indeed, for some purposes it may even be desirable to shade the houses still further. At certain times of year, however, the levels of radiation usually become limiting. Thus tomato growers in the U.K. encounter problems of poor flower development, pollination and setting during the early months of the year, and the quality of the fruit suffers. Under such conditions it is customary to use high levels of potassium and to attempt to restrict growth by limiting the water supply to the plants; high salinity levels are normally used in nutrient film culture at this stage.

In principle it is possible to supplement natural radiation levels with suitable electric lamps e.g. Boivin et al. (1984). In practice, however, this is usually too expensive for general use and is therefore confined to the propagating stage, when the plants can be grouped together at close spacing. In contrast to giving supplementary lighting at levels sufficient to provide effective amounts of energy for photosynthesis, control of flowering by adjustment of daylength is standard practice for year-round chrysanthemums. Covers of black polythene or cloth are used to shorten the day during summer, thus inducing flowering of this 'short-day' crop. Supplementary lighting at quite low intensities e.g. 100 lux, given during the early weeks of growth in winter, suffices to lengthen the day (shorten the night period) thus preventing premature budding on short stems. In general it is customary to reduce nitrogen applications during periods of low light and to increase the K:N ratio.

Enrichment with carbon dioxide. Green plants depend on photosynthesis for the continuing supply of organic compounds necessary for their growth. Under some conditions, particularly in the enclosed atmosphere of a greenhouse, the concentration of carbon dioxide in the air can become the limiting factor to carbon assimilation. The effects of increased levels of carbon dioxide on glasshouse crops were being studied at Cheshunt Research Station (U.K.) as long ago as 1922. Favourable crop responses were found, but the concept of aerial enrichment could not at that time be exploited commercially because of phytotoxic effects resulting from impurities in the fuels then available to produce carbon dioxide in the glasshouses. It was also at one time thought that enrichment with carbon dioxide would be favourable only during sunny weather, that is, when rates of photosynthesis were high. Gaastra (1964) showed, however, that in Holland a favourable response might be expected during some 80% of the days during the low-light month of January. Aerial enrichment with carbon dioxide, usually to about 1000 ml m^{-3}(vpm), is now standard practice for winter lettuce in the U.K. and N.W. Europe, and is essential for early tomato production in regions having poor winter light. The benefits and problems associated with carbon dioxide enrichment have been reviewed by Hand (1982). Increased production of dry matter resulting from supplementary carbon dioxide increases the uptake of nutrients by the crop.

The greenhouse as a planned production unit

As indicated in the preceding sections, the key feature of protected cropping lies in the ability to modify the environment in favour of crop growth and productivity. Not only can conditions for the crop be improved compared with those outdoors but the season can be extended considerably, in some cases to truly 'year-round' production, and crop range can be extended beyond that attainable locally in the open. The effects of wind damage, frost, hail or drought can be avoided, and the enclosed environment with easy access to the plants facilitates crop protection by chemical or biological methods.

As a consequence of all these advantages, protected cropping gives increased yields of high quality produce. This in turn requires increased nutrient uptake per unit area and the need for specialisation in planning the supply of nutrients throughout cropping. The high capital investment associated with greenhouse installations makes it essential that growth and yield are never limited by lack, excess or imbalance of the nutrients supplied to the crop. Such considerations place great responsibility on the grower, the adviser and the research scientist.

Fertiliser applications for agricultural and horticultural crops

Because of the far more favourable environment provided for greenhouse crops, with more rapid growth and, in some instances, extended cropping periods, it is only to be expected that the rates of fertiliser application far exceed those normally recommended in the field. Nevertheless, those who normally deal with field crops are frequently surprised by the heavy dressings given to greenhouse crops; the contrast is indeed very striking.

This point is illustrated by the data in Table 1 for phosphorus and potassium applications recommended in the U.K. for certain field vegetables and for glasshouse crops. The pre-planting applications are typical dressings as recommended when the results of a recent soil analysis are not available (MAFF, 1978, 1980a). The amounts of potassium supplied as top dressings to the glasshouse crops via irrigation equipment have been estimated on the basis of average water requirements and the recommended nutrient concentration of the liquid feeds. Although no estimates have been included for phosphorus in the liquid feeds, since the pre-planting application often suffices, supplementary phosphate is sometimes given. For example, it is not uncommon to include phosphorus at about 25 mg l^{-1} P for carnations, particularly in the second year of cropping. Inclusion of supplementary phosphorus would, of course, further accentuate the contrast in fertiliser usage between field and protected crops.

Situation	Crop		kg P ha^{-1}	kg K ha^{-1}
Field				
	Carrot		87	83
	Radish		26	104
	Brussels sprout		26	124
	Onion (bulb)		87	166
	Tomato (field)		26	166
Glasshouse				
	Carnation:	base fertiliser	308	558
		liquid feed*	–	912
	Tomato:	base fertiliser	205	1195
		liquid feed	–	1426
		total for season	205	2621

* for one year.

Table 1 Fertiliser recommendations for greenhouse and field crops where soil analyses are not available, together with the recommended top-dressing (liquid feeding) applications. Based on MAFF (1978, 1980a) and other sources.

The rates of potassium application for greenhouse tomatoes are particularly high, and the need for such dressings has sometimes been questioned by those more used to field crops. Somewhat lower levels would indeed suffice for vegetative growth and yield, but the higher rates are necessary to ensure high quality fruit (Roorda van Eysinga, 1966; Winsor and Long, 1967). It should, however, be emphasised that the fertiliser applications in Table 1 are for average situations; the recommendations

for soils initially of low nutrient status are far higher. For example, the potassium application suggested for tomatoes in a low-potassium soil rises to 1992 kg ha^{-1} of base fertiliser (MAFF, 1978). On a 'new' glasshouse soil it is also recommended that, for tomatoes, additional potassium sulphate should be incorporated below the normal cultivation depth i.e. in the 20–40 cm zone, at the rate of 1395 kg K ha^{-1}. These are indeed surprisingly high figures, nevertheless justified by the acknowledged difficulties of producing evenly-ripened tomato fruit in the first seasons of greenhouse production on a new site.

Forms of fertiliser used in the greenhouse

Despite the particularly heavy rates of fertiliser use with protected crops, the high value of the produce per unit area makes it possible to use more expensive forms where these offer special advantages. Whilst growers will rightly use the least expensive form of fertiliser suitable for their particular purpose, it is sometimes economic to apply relatively costly products where these either save labour or minimise the risk of damage to plants.

The early greenhouse growers tended to favour natural organic sources of nutrients. For example, nitrogen was supplied by a range of products such as horn, hoof, dried blood and guano, the rates of mineralisation of these increasing in the order listed. Horn and hoof were indeed capable of slow and sustained nitrogen release in soil, making them valuable for potting composts and for pre-planting application in glasshouse borders. Dried blood still attracts the older generation of gardeners as a relatively 'safe' nitrogen source having also the advantage of providing small doses of certain other nutrients e.g. iron.

Natural organic sources of nitrogen such as hoof and horn were generally unpleasant both to process and to handle, and were increasingly replaced by 'cleaner' slow-release fertilisers from about 1950 onwards. Thus some degree of slow release was attained by synthesising new chemical compounds, based on the reactions between urea and formaldehyde (Yee and Love, 1946). The resulting methylene-ureas varied in their rates of mineralisation in soil, depending on their chain-length (Long and Winsor, 1960) and the pH of the substrate (Winsor and Long, 1956). Analogous compounds such as crotonylidenediurea (CDU) and isobutylidenediurea (IBDU) also attracted attention in Germany and Japan respectively (Araten, 1968). Whilst such compounds all have their uses, none of them can really be regarded as an ideal slow-acting fertiliser; all are somewhat pH-dependent in their action, being hydrolysed more rapidly in acid than in neutral or calcareous soils. Furthermore, the proportion of truly 'slow-release' nitrogen is usually rather low; with ureaformaldehyde, for example, much of the nitrogen consists of either rapidly released or relatively unavailable nitrogen.

As an alternative to the synthesis of suitable chemical compounds to achieve slow mineralisation of nitrogen, physical restraints on nutrient release were also tried. Thus Oertli and Lunt (1962) studied the effects of certain surface coatings on the release of nutrients from fertilisers. Mixed 'NPK' fertilisers could be produced in this way, the products certainly providing a slow release of nutrients. Such fertilisers proved relatively expensive, however, and their use is usually restricted to special situations. One very successful application is in the peat or peat-sand substrates widely adopted for container-grown hardy ornamentals, the object being to minimise labour costs where the alternative would be to install and operate a liquid feeding system. Once a slow-release fertiliser has been applied, however, the grower virtually loses control over the nutrition of his crop. High temperatures, for example, can cause more rapid release of nutrients, leading to salinity problems and possible root damage. Furthermore, chemical analysis of soils and other substrates containing slow-release fertilisers can give rise to highly misleading results.

Application of nutrients in solution ('liquid feeding')

In striking contrast to the concept of slow-release fertilisers, the application of simple inorganic nutrients in solution whenever required, either via a hose or preferably by means of an irrigation system (see Black, 1976), has received increasing support in recent years. Watering greenhouse crops with an open hose is a time-consuming duty requiring great experience and is capable of destroying the structure of the surface soil. Irrigation systems make it possible to water large areas of greenhouses reproducibly at the turn of a tap.

Various types of irrigation systems are available e.g. MAFF, 1980b. Some such systems rely on the release of water via a screw thread at selected points along the supply line, whilst others use short lengths of small-bore ('spaghetti') tubing fed from a larger main to achieve the same effect. Separate outlets are usually provided for each plant e.g. tomatoes, cucumbers, though outlets may also be arranged on an area basis for closer-planted crops e.g. carnations, chrysanthemums. Low-level sprinkler systems are sometimes used for the same purpose, as also are stitched polyethylene film tubes and porous or 'leaky' hoses.

Irrigation has now become a relatively sophisticated procedure. For example, the various sectors of a large nursery will be irrigated sequentially, the timing and frequency of watering being pre-set and controlled automatically. The operator of such irrigation systems will normally have access to an information service giving the amount of water needed to restore the moisture deficit created during the previous day, based on the solar radiation levels. Modern nurseries with computerised control systems may be equipped to make such measurements on site, thus further simplifying the task of the supervisor. Nevertheless, some degree of judgement may still be required to meet the particular conditions of the crop.

Irrigation systems provide a most valuable opportunity to supply nutrients to crops throughout the whole period of their growth. No longer is it necessary either to rely entirely on the pre-planting fertilisers or to spread top-dressings of nutrients by hand, watered in by hose. Fertiliser applications during cropping can be tailored to meet the needs of the crop as growth progresses, giving an opportunity to vary the nutritional treatement at will according to the crop response. In contrast, the top-dressing of field crops is a labour-intensive process, and the choice of an appropriate base-dressing is something of an act of faith: the farmer has no advance knowledge of the environmental conditions which will prevail during cropping and has no means of influencing them. Heavy rainfall early in the season would leach away part of the fertiliser applied, whilst the crop might well have benefitted from additional nutrients during sustained periods of warm, moist weather. Predictions for field crops can only be based on typical 'average' conditions, whilst each year brings its own individual combination of environmental factors.

The forms of fertiliser used for liquid feeding are almost entirely inorganic, though urea has sometimes been advocated as a highly soluble nitrogen source. Whilst urea may be acceptable under some conditions e.g. in unsterilised soils at pH values below about 6.5, its use in sterilised media at higher pH could cause damage to plants owing to accumulation of ammonia-nitrogen (see p. 10).

The simple formulations used to provide nutrients in solution for soil-grown crops generally supply only nitrogen and potassium at K:N ratios ranging from up to 2:1 for tomatoes to nearer 1:1 for flower crops. The ingredients for such feeds would normally be potassium nitrate to supply the potasium and ammonium nitrate to give the additional nitrogen; see MAFF (1982).

For long-term crops e.g. carnations, or where a shortage of phosphorus might decrease yield, it is sometimes necessary to include phosphorus in the liquid feeds. In areas with hard water this can lead to precipitation of calcium phosphate and to the blocking of irrigation nozzles or capillary tubes. In such circumstances it can be helpful to switch to plain water in the final stages of each period of irrigation, thus clearing the nutrient solution from the outlets. Phosphorus is usually supplied as ammonium dihydrogen phosphate, though in hydroponic systems it is possible to provide part or all of the phosphorus requirement as phosphoric acid to maintain the required pH (~6.0) in the nutrient solution.

The choice of nitrogen source affects the pH of the soil or other substrate; ammonium salts lower the pH whilst nitrates raise it. Ammonium nitrate is better balanced in this respect, but its continuous use for long-term crops can nevertheless cause acidity. As an example of this, the average soil pH values in a nutritional study of carnations (Winsor and Adams, unpublished) are shown in Table 2 in relation to nitrogen application. Each plot, before planting, had received ground limestone at rates of 3–6 kg m^{-3}, together with calcium hydroxide at one tenth of these rates. The pH measurements, made 17 months after planting, show a progressive decrease with increasing nitrogen concentration.

N-concentration mg l^{-1}	Soil pH
100	7.0
140	6.6
180	6.4
220	6.2
260	6.0

Table 2 Soil pH after cropping with carnations for 17 months, during which period all water applied contained nitrogen, supplied as ammonium nitrate at the concentrations stated. Each value is the mean for eight plots.

In contrast, where potassium is supplied as potassium nitrate in the irrigation water and only the balance of the nitrogen requirement is provided as ammonium nitrate, the pH of the substrate will tend to rise, particularly at high K:N ratios and in areas having 'hard' water.

Magnesium is occasionally included in the liquid feeds for tomatoes, though this can only be regarded as supplementary to the use of magnesium sulphate as a pre-planting fertiliser application. If symptoms of magnesium deficiency develop, foliar sprays containing magnesium e.g. 1% $MgSO_4.7H_2O$, give more rapid control. Spraying is a time-consuming and therefore costly procedure, the need for which can be minimised by building up an adequate reserve of magnesium in the soil. Inclusion of trace elements in the irrigation water for soil-grown crops is not common, though carnations grown at pH 6.5 and above would in many cases benefit from boron supplied at 0.5 mg l^{-1} at every watering.

The distribution of nutrients by trickle ('drip') irrigation is by no means uniform. Certain nutrients e.g. nitrate, tend to move away from the outlets and concentrate in the drier areas of the soil, whilst more strongly adsorbed ions such as potassium remain in the region of the outlets (Winsor et al., 1961; Winsor and Long, 1962; Rolston et al., 1981). Little is known concerning the effects of this restricted distribution of nutrients through the root zone. It seems likely, however, that the tendency for potassium not to move freely away from the sites of irrigation would,

for example, lessen the effectiveness of high concentrations of this nutrient on tomato fruit quality when applied via drip nozzles; much of the water taken up by the crop could be drawn from areas of lower potassium concentration. If this is so, then the ratio of application of potassium in base and top dressings shown in Table 1 could with advantage be shifted towards the pre-planting treatment, since the latter is applied uniformly over the whole area.

Soil sterilisation

The restricted crop range in commercial greenhouses, often monoculture, together with the generally favourable environment of protected cropping, can lead to problems from soil-borne diseases. It was for this very reason that the first research station devoted to greenhouse crops in the U.K. (Experimental and Research Station, Cheshunt, Hertfordshire) was set up by growers in 1913. Partial sterilisation by steaming proved to be effective in maintaining high yields by eradication of pathogens, but the rising costs of both labour and fuel lead to powerful competition from chemical sterilants such as methyl bromide and metham-sodium. Chemical sterilisation can create problems of phytotoxicity, however, whereas soil steaming is fully acceptable as a routine procedure, even for 'year-round' crops such as chrysanthemums. Thus, within the sequence of harvesting and replanting, individual beds or groups of beds can be steamed to eliminate pathogens without injury to adjacent beds in which plants are growing. The choice between steaming and chemical sterilisation will always be governed by considerations of cost-effectiveness. Where chemical sterilisation suffices then this, being cheaper, is to be preferred; where the additional expense is justified, however, partial sterilisation by steaming will continue to be used.

Although steaming is not only widely effective but also avoids most problems of phytotoxicity, specific problems can arise under certain conditions. Thus, in acid soils, steaming can induce problems of manganese toxicity (e.g. Davies, 1957; Jager et al., 1969). The major glasshouse crops vary markedly in their susceptibility to excess manganese. Lettuce plants are particularly prone to sterilisation injury in steamed soils whereas tomatoes are relatively tolerant. Whilst lettuce plants may show leaf damage at concentrations of only 200–300 μg g^{-1} Mn (Messing, 1965; Roorda van Eysinga and Smilde, 1971) tomatoes are far less susceptible, concentrations of 2000 μg g^{-1} Mn or more being not uncommon in the leaves of plants grown in steamed soil. Such levels are unduly high for tomatoes and could readily be decreased by liming the soil to pH 6.8 or above, but the effect on yield would not necessarily be beneficial. Thus Winsor et al. (1967) reported lower yields in limed than in unlimed soil, despite visual symptoms of manganese toxicity in the unlimed plots. For further discussion and examples of toxicity symptoms in relation to leaf manganese content see Harrod (1971), Solbraa and Selmer-Olsen (1981).

Both steaming and chemical sterilisation of soils can temporarily suppress nitrification with a resultant accumulation of ammonium-N. Whilst this may be relatively harmless in moderately acid soils, problems can arise at higher pH, particularly where ammonium-fertilisers are used (see Winsor, 1964). The proportion of the total ammonium-N present as free ammonia increases rapidly with pH above about pH 7 (Court et al., 1964; Bates and Pinching, 1950) and it is this that causes damage to plants. For this reason it is suggested that urea, which releases ammonia rapidly by enzymatic breakdown, and heavy dressings of natural organic fertilisers e.g. hoof and horn (see Bunt, 1976) should not be used in limed soils after steaming.

Soilless culture

The intensive nature of glasshouse crop production makes great demands on the substrate with regard to physical properties. Soils which give adequate growth of plants in the field would, nevertheless, rarely be ideal as a medium for glasshouse crops in pots, beds, or borders. Thus, for maximum returns from the costly facilities of protected cropping, it is necessary to have high moisture retention without loss of aeration.

Natural organic substrates. Materials such as peat, provided they are not excessively humified i.e. not above index 3–5 on the von Post scale (Puustjarvi, 1970), meet these requirements admirably. A peat suitable for horticultural use would have about 94–95% of pore space, thus offering possibilities of good aeration and water retention. For example, a sample of Irish sphagnum peat, when thoroughly wetted and drained, had 17.1% air-filled porosity and an 'easily available' water content of 31% (Bunt, 1984). Peat, either alone or mixed with sand, vermiculite or perlite is thus an excellent and widely used substrate for crop production in pots, bags and beds; for further details see Bunt (1976) and Bauerle (1984).

Because of the high cost of peat, attention has also been given to other natural organic materials, particularly as substrates for use in pots or bags. Sawdust has been used extensively in Canada (Maas and Adamson, 1980; Adamson and Mass, 1981) and elsewhere (Maree, 1981, 1984). Fresh and composted wood bark have also been tested and used for crop production e.g. Verdonck et al. (1983) and Pudelski (1983). From a nutritional standpoint, the nitrogen supply to these substrates needs careful control relative to microbiological transformations. Problems have also arisen in some instances due to manganese toxicity, since this element is often present in the bark at quite high concentrations (Solbraa and Selmer-Olsen, 1981).

Though soilless substrates provide an excellent physical medium for plant roots, they usually provide little, if any,

of the nutrients required for plant growth and, when fertilised, are prone to nutrient losses by leaching. Whereas part of the potassium in soils is held in exchangeable form, most of this element in a fertilised peat remains water-soluble and contributes markedly to the total salinity (Adams et al., 1978b). Nutrients such as phosphorus and potassium are far more prone to leaching from peats than from soils, and it is more often necessary to include phosphorus in the nutrient solutions ('liquid feeds') applied to peat than to soils. Peat, though relatively resistant to further decomposition, is by no means an inert substrate. Thus fresh, relatively unhumified, peat will immobilise considerable amounts of added nitrogen by microbial action; crops grown in 'new' peat may appear somewhat low in nitrogen despite the inclusion of this element both in base dressings and in liquid feeds. At a later stage, however, usually in the second season of cropping, this process will be reversed (Adams et al., 1978). Nitrogen will be released from the peat by mineralisation, and salinity levels will rise accordingly. It is therefore important to supply adequate amounts of nitrogen in the first season of cropping but to avoid high salinity levels in the following season by restricting the pre-planting application of this nutrient.

Deficiencies of micronutrients are not particularly common in soil-grown glasshouse crops, though the problems of certain combinations of crops and mineral deficiencies are well known e.g. boron deficiency of carnations in limed soil, resulting in splitting of the calyx (Winsor et al., 1970). In soilless substrates it is usually necessary to supply all the essential elements, both the macronutrients and the micronutrients.

The most convenient method of providing the full range of micronutrients is as a fritted trace element mixture (see Bunt, 1976). Fritted materials have a valuable 'slow-release' characteristic, but the availability of micronutrients to the plant is still greatly influenced by the pH of the substrate. Thus the supply of molybdenum from a frit used in an acidic substrate may prove inadequate for certain crops such as greenhouse lettuce. Similarly, boron deficiency can occur in a limed substrate after the first few months of cropping, the frit being unable to maintain an adequate supply of boron to the plants under unfavourable conditions.

Peat has the ability to 'fix' copper, making it unavailable to plants, and it is often easier to demonstrate deficiencies of this element in peat than in sand or solution culture. Examples of greenhouse crops showing copper deficiency when grown in peat without added copper include cucumbers, for which leaf chlorosis and reduced yields have been reported (Adams et al., 1978a) and chrysanthemums, which fail to flower when deprived of copper (Adams et al., 1975).

Hydroponics. Numerous techniques have been devised for growing plants in nutrient solutions, with or without an inert substrate e.g. sand or gravel present. Such techniques have proved invaluable in the study of plant nutrition, their use extending back at least to the early nineteenth century e.g. De Saussure, 1804. Soilless culture has played an important role in establishing the essential nature of various elements for the growth and reproduction of plants, using criteria proposed by Arnon and Stout (1939). Again, and on a wider scale, sand and water culture methods are much used to demonstrate not only the visual symptoms of mineral deficiencies and toxicities but also quantitative growth responses to nutrient concentration.

The techniques of sand and water culture, including particular attention to the purity of the chemicals, water and substrate (if any) used, have been described comprehensively by Hewitt (1966), and will not be discussed further here. Mention should, however, be made of commercial systems of crop production in soilless culture, and particularly of the more recent system known as the nutrient film technique ('NFT'); for general descriptions of this latter system see Spensley et al., 1978; Winsor, 1980.

Nutrient film culture (Cooper, 1975) is a system using no solid substrate other than the small pot or block of synthetic material in which the plants were raised. The distinguishing feature of 'NFT' is the extremely shallow layer of nutrient solution used, preferably only a few millimetres in depth. The solution flows down plastic gullies, in which the crop is grown, to a catchment tank where it is dosed automatically with nutrients on the basis of a pre-set salinity (electrical conductivity) level and with acid (nitric and/or phosphoric) to maintain the pH at about 6.0. The solution is then pumped to the tops of the gullies to complete the cycle; for a detailed description see Winsor et al. (1979).

Crop production in soilless culture systems such as nutrient film culture requires that an adequate supply of all the elements essential for plant growth be maintained in solution. This is achieved by a combination of automated salinity control, periodic chemical analyses and changes in the formulation of the fertiliser concentrates. Being a 'closed' system, unlike some systems in which unused nutrient solution runs to waste, it is important that the fertilisers used do not supply excessive amounts of ions such as chloride or sulphate which would accumulate in the system. The ratio of the nutrients provided in the fertilisers and water supply should, as far as possible, match their uptake by the crop. If this is not achieved, certain ions will accumulate progressively in the solution; the salinity controller will then display a seemingly adequate conductivity level and injection of further nutrients will be prevented, resulting in depletion of various essential elements. Certain constituents of the water supply may accumulate markedly in a closed system. Thus very high levels of calcium ions may build up in areas having 'hard' water. High levels of sodium and chloride

can also occur, and zinc can accumulate to excess where galvanised iron pipes are used in the nursery water supply.

In general, relatively low concentrations of nutrients suffice for crop growth in flowing solution culture. In contrast to static systems, the plant roots are bathed in a solution of maintained concentration, and nutrient uptake is not greatly restricted by concentration gradients at the root surface. No significant differences in the yields of tomato fruit were found over the range 10–320 mg l^{-1} NO_3–N (Massey and Winsor, 1980a). Similarly, the yield of tomatoes was not affected significantly by phosphorus concentration over the range 5–200 mg l^{-1} P (Massey and Winsor, 1980b). Not all crops are unaffected by high phosphorus concentration, however. Thus Asher and Loneragan (1967), working with 8 Australian pasture species in flowing nutrient solutions, reported mild phosphorus toxicity in 3 species at only 0.16 mg l^{-1} P, becoming severe at 0.74 mg l^{-1}. Similarly, severe toxicity symptoms developed in Lupinus luteus at 2.5 mg l^{-1} P in nutrient film culture (Winsor and Grimmett, 1984). Some plants are thus very sensitive to phosphorus supply, accumulating high concentrations of this element in their tissue from surprisingly low concentrations in the culture solution.

Synthetic substrates such as 'rockwool' (see Verwer, 1975, 1978; MAFF, 1984) have also attracted much attention for commercial crop production. Apart from the presence of a solid substrate throughout the rooting zone, systems based on such materials differ from the nutrient film technique in that the nutrient solution is not normally recirculated. The rockwool slabs are enclosed with polythene sheet but are not arranged on a sloping surface; any excess nutrient solution, supplied by a conventional glasshouse irrigation system terminating in small bore plastic tubes, flows slowly to waste. The equipment for controlling pH and nutrient concentration is broadly similar to that used for NFT. Thus, two containers are required for stock solutions of fertilisers and a third for dilute acid. The sensors for pH and conductivity monitor and control the injection of concentrates into the mixing tank, and the resulting solution is pumped into the irrigation system. Simple nursery diluters are not adequate for this purpose.

Rockwool culture is now widely used in the Netherlands and elsewhere in N.W. Europe, particularly for cucumbers, tomatoes and peppers. Being a non-recirculating or 'open' system, it is regarded by many growers as somewhat simpler to control than NFT. The nutrient solution in the slabs needs to be sampled and analysed periodically and the nutrient supply is adjusted accordingly. The technique is somewhat similar to that of growing in peat bags, and thus forms a natural progression for many growers, substituting an inorganic for an organic substrate. Mineral wool is relatively expensive to produce and has to be sterilised between crops and replaced at intervals of 2–3 years. Experience with properly managed NFT systems has shown that the presence of a substrate is in no way essential for commercial tomato production. It would therefore seem only logical for the continuing development and adoption of hydroponic techniques to progress ultimately towards systems not involving any solid substrate.

Layout and content of the chapters which follow

The text and illustrations cover seven glasshouse crops, starting with the major vegetable crops and continuing with flowers. Within each of these main groupings the crops are arranged in alphabetical order. For each crop the text starts with a literature review and bibliography; colour plates with descriptive notes are presented in one section following the last chapter and they follow the same crop sequence as do the chapters. The literature reviews for each crop are sub-headed in a regular sequence of macronutrients and micronutrients, and for each nutrient the text deals firstly with deficiency symptoms and crop responses followed by analytical data for the element concerned. Toxicities are also referred to where relevant to glasshouse as distinct from field crops. Only limited data have been included for sulphur, since the customary inclusion of sulphate-containing fertilisers e.g. potassium and magnesium sulphates and superphosphate, makes deficiency of this element virtually unknown in commercial glasshouses. Except where stated otherwise, all plant analyses have been presented on the oven-dry weight basis.

It should be emphasised that the descriptions and analytical values in the literature reviews are based on data published throughout the world. In contrast, the colour plates (with six exceptions), the descriptive captions and the analytical data included therein are from nutritional studies at the Glasshouse Crops Research Institute (GCRI). Most of the illustrations are from the extensive collection of such photographs taken in the course of long-term nutritional studies at this Institute. The early studies were of crops grown in glasshouse border soils, followed by experiments in peat and peat-sand substrates and later in recirculating solution culture (NFT). The object throughout has been to illustrate deficiency or toxicity symptoms and nutrient imbalances for mature plants grown under quasi-commercial conditions. Where necessary, however, disorders which had not already occurred in the course of our main nutritional studies were specifically induced in pot-grown plants. Results so obtained are usually very striking, though sometimes less immediately recognisable by commercial growers.

The descriptive notes accompanying the colour plates include, where possible, analyses of the actual or comparable leaf tissue. The ranges of analytical values which we generally associate with normal growth, deficiency or excess nutrient levels are also indicated. The ranges for normal growth give the approximate limits within which

no very obvious visual indications of either deficiency or excess would normally be expected under the conditions of our studies. It is not, however, implied that the yield or quality would necessarily be equally satisfactory throughout the entire range; the optimum probably lies near the middle of the quoted range in every case.

The generalised analytical values given with the plates apply to the standardised though arbitrarily chosen sampling positions widely adopted, namely the young but almost fully expanded ('young mature') leaves. The sampling positions used at GCRI are as follows:–

Tomato:	fifth separated leaf below the head of the plant.
Cucumber:	fifth or sixth leaf below the head.
Carnation:	fifth pair of leaves from the top of a vegetative shoot.
Chrysanthemum:	seventh leaf from the top.
Lettuce:	segments of whole lettuce heads, excluding any soiled or senescent outer leaves.

References

Adams, P., Graves, C. J. and Winsor, G. W. (1975). Some effects of copper and boron deficiencies on the growth and flowering of *Chrysanthemum morifolium* (cv. Hurricane). *J Sci Fd Agric* **26**, 1899–1909.

Adams, P., Graves, C. J. and Winsor, G. W. (1978a). Effects of copper deficiency and liming on the yield, quality and copper status of tomatoes, lettuce and cucumbers grown in peat. *Scientia Hort* **9**, 199–205.

Adams, P., Graves, C. J. and Winsor, G. W. (1978b). Tomato yields in relation to the nitrogen, potassium and magnesium status of the plants and of the peat substrate. *Pl Soil* **49**, 137–148.

Adamson, R. M. and Maas, E. F. (1981). Soilless culture of seedless greenhouse cucumbers and sequence cropping. *Agriculture Canada (Ottawa)*, Publication **1725/E**, 21 pp.

Araten, Y. (1968). New fertiliser materials. *New Jer, Noyes Dev Corpn.*

Arnon, D. I. and Stout, P. R. (1939). The essentiality of certain elements in minute quantity for plants with special reference to copper. *Pl Physiol* **14**, 371–375.

Asher, C. J. and Lonergan, J. F. (1967). Response of plants to phosphate concentration in solution culture. I. Growth and phosphorus content. *Soil Sci* **103**, 225–233.

Bates, R. G. and Pinching, G. D. (1950). Dissociation constant of aqueous ammonia at 0 to 50 from E.M.F. studies of the ammonium salt of a weak acid. *J Am chem Soc* **72**, 1393–1396.

Bauerle, W. L. (1984). Bag culture production of greenhouse tomatoes. *Ohio Agricultural Research and Development Centre*, Special Circular *108*, 7 pp.

Black, J. D. F. (1976). Trickle irrigation – a review. *Hort Abstr* **46**, 1–7 and 69–74.

Boivin, C., Gosselin, A. and Trudel, M. J. (1984). Potentiel de l'eclairage artificiel pour la production de tomates de serre. Symposium Internationale sur la Serriculture, 26–27 Octobre 1984. *Université Laval, Quebec*, 69–90.

Bunt, A. C. (1976). Modern potting composts: a manual on the preparation and use of growing media for pot plants. 277 pp. *Lond, Allen & Unwin*.

Bunt, A. C. (1984). Physical properties of mixtures of peats and minerals of different particle size and bulk density for potting substrates. *Acta Hort* **150**, 143–153.

Cooper, A. J. (1975). Crop production in recirculating nutrient solution. *Scientia Hort* **3**, 251–258.

Court, M. N., Stephen, R. C. and Waid, J. S. (1964). Toxicity as a cause of the inefficiency of urea as a fertiliser. Parts I and II. *J Soil Sci* **15**, 42–65.

Davies, J. N. (1957). Steam sterilisation studies. *Rep Glasshouse Crops Res Inst 1954–1955*, 70–79.

Gaastra, P. (1964). Fysiologische aspecten van de toepassing van koolzuur. *Meded Dir Tuinb* **27**, 369–377.

Hand, D. W. (1982). CO_2 enrichment, the benefits and problems. *Scient Hort* **33**, 14–43.

Harrod, M. F. (1971). Metal toxicities in glasshouse crops: discussion of problems encountered in advisory work on soils of pH 6.0 and above. *Minist Agric Fish Fd Tech Bull* **21**, 176–192. Lond HMSO.

Hewitt, E. J. (1966). Sand and water culture methods used in the study of plant nutrition. *Tech Commun Commonw Bur Hort Plantn Crops no.* 22, 2nd ed, 547 pp.

Jager, G., van der Boon, J. and Rauw, G. J. G. (1969). The influence of soil steaming on some properties of the soil and on the growth and heading of winter glasshouse lettuce. I. Changes in chemical and physical properties. *Neth J agric Sci* **17**, 143–152.

Lawrence, W. J. C. (1963). Science and the glasshouse (3rd edit). *Edin, Oliver & Boyd*, 183 pp.

Long, M. I. E. and Winsor, G. W. (1960). Isolation of some urea-formaldehyde compounds and their decomposition in soil. *J Sci Fd Agric* **11**, 441–445.

Maas, E. F. and Adamson, R. M. (1980). Soilless culture of commercial greenhouse tomatoes. *Agriculture Canada (Ottawa)*, Publication **1460**, 27 pp.

Maree, P. C. J. (1981). Growing greenhouse cucumbers in fresh sawdust. *Dept Agron Past, Univ Stellenbosch* 12 pp.

Maree, P. C. J. (1984). Growing seedless English cucumbers in fresh pine sawdust and bark. *Proc 6th Int Cong on Soilless Culture, Lunteren*, 355–363. *Wageningen, ISOSC.*

Massey, D. M. and Winsor, G. W. (1980a). Some responses of tomatoes to nitrogen in recirculating solutions. *Acta Hort* **98**, 127–137.

Massey, D. M. and Winsor, G. W. (1980b). Some responses of tomato plants to phosphorus concentration in nutrient film culture. *Proc 5th Int Cong on Soilless Culture Wageningen*, 205–214.

Messing, J. H. L. (1965). The effects of lime and superphosphate on manganese toxicity in steam-sterilised soil. *Pl Soil* **23**, 1–16.

Ministry of Agriculture, Fisheries and Food (1978). Lime and fertiliser recommendations No. 4: Glasshouse crops and nursery stock, **GF 24**, 39 pp.

Ministry of Agriculture, Fisheries and Food (1980a). Lime and fertiliser recommendations No. 2: Vegetables and bulbs, Booklet **2192**, 20 pp.

Ministry of Agriculture, Fisheries and Food (1980b). Watering equipment for greenhouse crops, Booklet **2140**, 26 pp.

Ministry of Agriculture, Fisheries and Food (1982). Lime and fertiliser recommendations No. 4: *Glasshouse crops and nursery stock 1983/84. Booklet* **2194**, 48 pp.

Ministry of Agriculture, Fisheries and Food (1984). Tomato production 6: Hydroponic growing systems. Booklet **2249**, 43 pp.

Moorby, J., Graves, C. J. and Fish, V. S. (1979). The effect of root temperature on tomato growth. *Rep Glasshouse Crops Res Inst 1978*, 85–86.

Moorby, J., Slough, P. C. and Fish, V. S. (1977). Effects of root temperature on growth of tomatoes. *Rep Glasshouse Crops Res Inst 1976*, 76–77.

Oertli, J. J. and Lunt, O. R. (1962). Controlled release of fertiliser minerals by incapsulating membranes. I. Factors influencing the rate of release. *Proc Soil Sci Soc Am* **26**, 579–583.

Pudelski, T. (1983). Composted and non-composted woodwastes in growing vegetables under protection in Poland. *Acta Hort* **133**, 237–240.

Puustjarvi, V. (1970). Degree of decomposition. *Peat and Pl News* **3(4)**, 48–52.

Rolston, D. E., Rauschkolb, R. S., Phene, C. J., Miller, R. J., Uriu, K., Carlson, R. M. and Henderson, D. W. (1981). Applying nutrients and other chemicals to trickle-irrigated crops. *Div Agric Scis Univ Calif, Bull* **1893**, 14 pp.

Roorda van Eysinga, J. P. N. L. (1966). Bemesting van tomaten met kali. *Publties Proefstn Groenten-en Fruitteelt Glas Naaldwijk* No. **109**, 37 pp.

Roorda van Eysinga, J. P. N. L. and Smilde, K. W. (1971). 'Nutritional disorders' in glasshouse lettuce. *Centre for agric Publ Documen, Wageningen*, 56 pp.

De Saussure, T. (1804). Recherches chimiques sur la végétation. Paris.

Smith, E. M. and Treaster, S. A. (1981). Evaluating hardwood and pinebark media for container grown woody ornamentals. *Ohio Agric Res and Dev Centre Res Circ* **263**, 3–6.

Solbraa, K. and Selmer-Olsen, A. R. (1981). Manganese toxicity – in particular when growing plants in bark compost. *Acta Agric Scand* **31**, 29–39.

Spensley, K., Winsor, G. W. and Cooper, A. J. (1978). Nutrient film technique – crop culture in flowing nutrient solution. *Outlook on Agric* **9**, 299–305.

Verdonck, O., De Vleeschauwer, D. and Penninck, R. (1983). Bark compost, a new accepted growing medium for plants. *Acta Hort* **133**, 221–226.

Verwer, F. L. J. A. W. (1975). Growing vegetable crops in rockwool and other media. *Acta Hort* **50**, 61–67.

Verwer, F. L. J. A. W. (1978). Research and results with horticultural crops grown in rockwool and nutrient film. *Acta Hort* **82**, 141–147.

Winsor, G. W. (1964). Some glasshouse practices. *Rep Prog appl Chem* **49**, 326–331.

Winsor, G. W. (1968). The nutrition of glasshouse and other horticultural crops. *Proc Fert Soc* **103**, 55 pp.

Winsor, G. W. (1980). Progress in nutrient film culture. *Span* **23(1)**, 7–9.

Winsor, G. W. and Grimmett, Molly M. (1984). Phosphorus toxicity. *Rep Glasshouse Crops Res Inst 1982*, 58–62.

Winsor, G. W. and Long, M. I. E. (1956). Mineralisation of the nitrogen of urea–formaldehyde compounds in relation to soil pH. *J Sci Fd Agric* **7**, 560–564.

Winsor, G. W. and Long, M. I. E. (1962). A factorial nutritional study of the perpetual-flowering carnation grown under glass. *J hort Sci* **37**, 299–312.

Winsor, G. W. and Long, M. I. E. (1967). The effects of nitrogen, phosphorus, potassium, magnesium and lime in factorial combination on ripening disorders of glasshouse tomatoes. *J hort Sci* **42**, 391–402.

Winsor, G. W., Davies, J. N. and Long, M. I. E. (1961). Liquid feeding of glasshouse tomatoes; the effects of potassium concentration on fruit quality and yield. *J hort Sci* **36**, 254–267.

Winsor, G. W., Davies, J. N. and Long, M. I. E. (1967). The effects of nitrogen, phosphorus, potassium, magnesium and lime in factorial combination on the yields of glasshouse tomatoes. *J hort Sci* **42**, 277–288.

Winsor, G. W., Hurd, R. G. and Price, D. (1979). Growers' Bulletin No. 5; Nutrient Film Technique. *Littlehampton, Glasshouse Crops Res Inst*, 47 pp.

Winsor, G. W., Long, M. I. E. and Hart, B. M. A. (1970). The nutrition of glasshouse carnations. *J. hort Sci* **45**, 401–413.

Yee, J. W. and Love, K. S. (1946). Nitrification of urea-formaldehyde reaction products. *Proc Soil Sci Soc Amer* **11**, 389–392.

Cucumber

There are many reports on nutritional trials with cucumber, (*Cucumis sativus L.*) but the majority of these are for field crops or pickling cucumbers. Of those reported for crops grown under glass, many are concerned with the responses of young plants and few with mature (fruiting) crops.

Unlike tomatoes, cucumbers are very sensitive to salinity. Thus yields declined as the electrical conductivity of the soil extracts increased (r = −0.95; Sonneveld and Beusekom, 1974) and as the electrical conductivity of the irrigation water increased (r = −0.90; Sonneveld and Voogt, 1978; Sonneveld, 1981). The regression of 'relative yield' (y), expressed as a percentage, on the electrical conductivity of the irrigation water (x, in mS cm^{-1} at 25°C) was given by the equation y = −16.8 x + 115. Little difference was found between individual salts in their depression of yield except for sodium bicarbonate, which raised the pH and affected soil structure adversely, and for magnesium chloride which caused excessive uptake of magnesium. Sonneveld (1981) proposed upper limits of 30 mg l^{-1} Na, 50 mg l^{-1} Cl and an electrical conductivity not exceeding 0.5 mS cm^{-1} (25°C) in the irrigation water. Water containing up to twice these levels could be used, however, if additional irrigation were given to cause leaching of the soil.

Nitrogen

Symptoms of deficiency and excess. As with most crops, the typical symptoms of N deficiency in cucumbers are yellowing of the foliage and restricted growth. Dearborn (1936) reported that cucumber plants grown at a low N level had stiff, woody stems and pale yellow-green foliage. Growth was reduced by the deficiency, and the fruits were of poor market quality, pale in colour and constricted at the distal end. Skinner (1941) described the deficiency in terms of stunted growth and yellow-green leaves, becoming yellow in extreme cases. The stems were thin and the fruit pale in colour. Poor development of the tip of the fruit was illustrated in colour, based on the work of I.C. Hoffman at Ohio Agricultural Experimental Station. Roorda van Eysinga and Smilde (1969, 1981) noted that the fruits from deficient plants (cv. Sporu) were short, pale and somewhat spiny whereas normal fruits were smooth and dark green. Paleness of deficient fruit was also illustrated by Wetzold (1972).

The symptoms of N excess have been described and illustrated by Roorda van Eysinga and Smilde (1969,

1981), Bergmann *et al.* (1965) and Bergmann (1976). The plants are very dark green, the leaves tend to curl downwards and the petioles droop slightly. Mild toxicity can cause a narrow band of marginal chlorosis, whereas severe toxicity leads to marginal and interveinal chlorosis and leaf scorch. Hanada *et al.* (1981) grew cucumber plants in solution culture over the range 0-896 mg l^{-1} N (64 me l^{-1}); toxicity symptoms developed at the highest N concentration, including leaf scorch and depression of growth.

Growth responses. Changes in the N status of cucumber plants are reflected rapidly in both the growth and the colour of the new leaves. Marked increases in growth were found in response to increasing levels of NO$_3$-N over the range 3-84 mg l^{-1} N (El-Sheikh and Broyer, 1970), though growth was depressed significantly at very high levels of N (672 mg l^{-1} NO$_3$-N). Barker and Maynard (1972) observed a significant reduction in the fresh and dry weights of plants at 280 mg l^{-1} NO$_3$-N. The latter authors also included treatments supplying different forms and proportions of inorganic N at 15 me l^{-1} N (210 mg l^{-1} N). Growth (dry weight) was highest with NO$_3$-N, and was reduced by 19% and 35% respectively when one third and two thirds of the total N was supplied in the form of NH$_4$-N. The poorest growth of all (64% reduction) occurred when all the N was supplied as NH$_4$-N. Schenk and Wehrmann (1979a) observed a small increase in the growth of young cucumber plants when NH$_4$-N (5 and 10 m M) was added to an 'all-nitrate' culture solution (3 m M NO$_3$) at pH 6.5. Raising the pH of the solutions, however, thus increasing the concentration of NH$_3$-N, decreased growth progressively and was accompanied by a reduction in chlorophyll content of the leaves. The early symptoms of ammonia injury were small chlorotic spots on the leaves; these subsequently increased in size and merged, leaving only the veins green.

Restriction of growth by inadequate levels of N, noted within four days from planting in beds of peat, was reflected in both the early and final yields of fruit (Adams *et al.*, 1974a). Early yields were improved by the highest concentration of 300 mg l^{-1} N in the liquid feed but, used at every watering, this subsequently became excessive and depressed the final yield by 10%. Further work suggested that a concentration of 140 mg l^{-1} N was fully adequate (Adams *et al.*, 1977). These results were obtained in beds containing 0.11 m^3 of peat per plant. Since the volume of

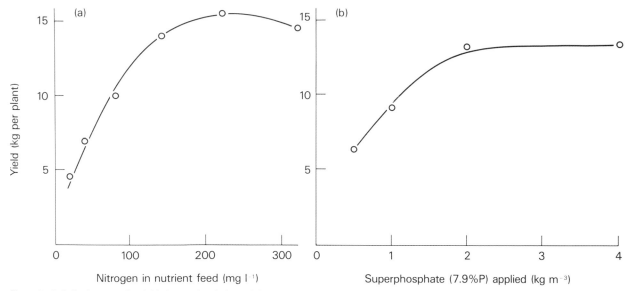

Figure 1 Relation between the yield of cucumber fruit and (a) the concentration of nitrogen in the liquid feed (after Adams *et al.*, 1978a) and (b) the amount of superphosphate applied before planting (after Adams, 1978).

peat per plant is an important factor in determining the amount of solution retained after each watering, plants grown with less peat may require correspondingly higher concentrations of nutrients in the liquid feed. The response to N was therefore studied in beds containing 0.04 m^3 of peat per plant which received a wide range of N concentrations (20–320 mg l^{-1} N) in the feed (Adams *et al.*, 1978a). Under these conditions the highest yield was obtained when the liquid feed contained 220 mg l^{-1} N (see Fig. 1a). This response was established during the first two weeks of harvesting and remained consistent throughout cropping. The proportion of poorly developed fruit increased to over 25% where the plants received less than 100 mg l^{-1} N in the liquid feed, and over 40% of the fruit were downgraded for poor colour when the feed contained only 20–40 mg l^{-1} N.

Light intensity (MJ m^{-2} d^{-1})	Uptake per plant		
	Water (l)	N (mg)	K (mg)
2.3	0.51	154	136
15.5	1.56	257	325
19.2	2.14	260	354

Table 3. Effect of solar radiation on the daily uptake of water, nitrogen and potassium by cucumber plants.

Massey (1978) grew cucumbers with a wide range of N concentrations (10–320 mg l^{-1} N) in recirculating solution culture. In the early stages of growth, plants receiving the lowest concentrations were palest whilst those receiving the highest concentration were darkest and many of the leaves developed a marginal scorch. However, this difference between treatments subsequently disappeared and

there was no significant difference between the yields. Uptake was closely related to light intensity (Table 3) and air temperature (Table 4) and highly correlated with water usage (r = 0.95; Adams, 1980).

x variates	y variate	
	Water, nitrogen or potassium	Phosphorus
	10 observations	7 observations
Light	63–76 (r = 0.82–0.89[xx])	49 (r = 0.756[x])
Air temperature	58–65 (r = 0.789–0.83[xx])	64 (r = 0.84[xx])
Solution temperature	34–49 (r = 0.64–0.74[x])	83 (r = 0.93[xx])
Light + air temperature + solution temperature	73–83	82
Light + air temperature	72–85	56
Light + solution temperature	77–84	79
Air temperature + solution temperature	53–66	84

Table 4. The proportion (%) of the variance accounted for by regressions relating the uptake of water and nutrients by cucumber plants to light intensity, air temperature and solution temperature. Since the values for water, nitrogen and potassium are very similar, only the ranges are given for these three variates. (after Adams, 1980).

Nitrogen content of the leaves. The total N content of the laminae is frequently used to assess the N status of the crop. Ward (1967) showed that this could vary from 10% in the oldest leaves to 6.7% N in the youngest; young mature leaves contained 5.5–6.0% N. A 'reference standard' of 6.0% N in cucumber laminae was later suggested by Ward (1973a), the sampling position chosen being the third visible leaf from the top of the stem i.e. a relatively young leaf, described as about 10 cm in diameter. Other authors have associated rather lower N contents with the best growth e.g. 4.1% N (Geissler, 1957, Barker and Maynard 1972), and with good yields 4.5–5.0% N (Adams, 1978); the difference may in some cases be due to the use of a rather lower leaf-sampling position. Thus Wetzold (1972) recorded 5–7% N in the younger leaves but only 2.5–3.5% N in the older leaves; the corresponding levels for deficient plants were <3.0% and <2.0%. Lepiksaar (1965) associated less than 1.75% N with a deficiency; Roorda van Eysinga (1970) reported 2.1–3.8% N in the leaves from glasshouse trials at Naaldwijk; Cheng and Forest (1977) quoted 4.3% N as a 'diagnostic' value for the mature leaves of field-grown cucumbers.

The NO_3–N content of the leaves is a good indicator of the N status of the crop; healthy plants contain 0.1–1.6% NO_3–N in the leaf dry matter whereas deficient plants contain 0.1% NO_3–N or less (Roorda van Eysinga and Smilde, 1969). Sonneveld (1981) proposed a 'guide value' of 0.5% NO_3–N in young, fully developed cucumber leaves.

Nitrogen levels in cucumber petioles were examined by El-Sheikh et al. (1970), El-Sheikh and Broyer (1970), the plants being grown in sand culture over the range 2.6–672 mg l^{-1} N. Maximum growth (dry matter production) corresponded to 2.1–3.4% N (total) and 0.57–1.19% NO_3–N in the mature petioles: correspondingly lower NO_3–N levels were recorded in young petioles (0.38–0.98%) and higher levels in 'old' petioles (0.68–1.49%).

Phosphorus

Deficiency symptoms. The growth of cucumber plants receiving an inadequate supply of P is restricted, although the leaves may show no definite symptoms (Roorda van Eysinga and Smilde, 1969). With severe deficiency, however, the plants are stunted and the younger leaves become small, stiff and dark greyish-green. Large brown areas, which include both veins and interveinal areas, develop on the lower leaves and spread up the plant. The affected leaves quite rapidly become completely shrivelled (Roorda van Eysinga and Smilde, 1969). Wetzold (1972) showed a generally poor leaf colour in P deficient plants grown in sand culture, with interveinal and marginal chlorosis and desiccation. The description of P deficiency given for young plants in solution culture by Holcomb and Hickman (1979) included reduction of leaf size, somewhat stunted growth, leaf scorch and subsequent drying of the tissue.

Growth responses. The yield response to increasing amounts of fertiliser P, applied to peat which was

maintained at pH 6, is shown in Figure 1b. When the peat was limed more heavily to maintain pH 7, the average yield was reduced by 10% (Adams, 1978). Phosphorus deficiency increased the proportion of poorly developed fruit, particularly in limed peat (Adams and Winsor, 1984). The yield of field-grown cucumbers more than trebled over the range 0-1760 kg ha^{-1} P, but decreased by 38% at a still higher rate of application (3520 kg ha^{-1}; Jaszczolt and Nowosielski, 1974).

In sand culture the growth (fresh weight) of cucumber plants increased by an average of 34% over the range 35 to 70 mg l^{-1} P, the corresponding contents of the leaves being 0.45 and 0.84% P in the dry matter; there was little or no further increase in growth at 105 mg l^{-1} P (Cheng and Forest, 1977). In nutrient-film culture, concentrations of 3–243 mg P l^{-1} in the recirculating solution had little effect on the early yield, though the highest concentration (243 mg l^{-1}) ultimately reduced both the weight and numbers of fruit harvested (Massey et al., 1983). The uptake of P appeared to be closely related to the temperature of the recirculating nutrient solution; light intensity and air temperature had less direct influence than on the uptake of N and K (see Table 4; Adams, 1980).

Phosphorus content of the leaves. Considerable data on the P content of cucumber leaves have been published. Geissler (1957) reported 0.7% P in the young foliage, falling to 0.35% in the older leaves. Wetzold (1972) found rather higher values of 0.8–1.5% P in the young leaves, again decreasing somewhat to 0.6–1.3% P in the older foliage. Phosphorus deficient plants contain less than 0.3% and 0.2% P respectively in young and old leaves. Ward (1967) made a detailed study of the distribution of P in cucumber plants, including leaves both on the main stem (positions 1–26) and on the laterals. Although the pattern of distribution was not entirely consistent, with few exceptions the values tended to be higher at the top of the plant. The range of values recorded was 0.63–1.36% P for main-stem leaves and 0.73–1.33% P for lateral leaves; the average for all leaf tissue on the plants, representing 46% of the total plant dry matter, was 0.91% P. Roorda van Eysinga and Smilde (1969) recorded 0.35–0.74% P in the leaves of healthy plants, and sometimes up to 1.0%, whereas the leaves of deficient plants contained only 0.13% P. Adams (1978) found that 0.7–1.0% P in the leaves was associated with good yields. Ward (1973a) suggested that a young leaf of about 10 cm diameter, usually the third leaf visible below the top of the plant, should be taken for plant analysis and a reference standard of 1.0% P was proposed. Sonneveld (1981) quoted a far lower 'guide value' of 0.5% P in the 'young, fully developed leaves'.

Further values are available for cucumber plants grown in the field. Thus Bradley et al. (1961), sampling the oldest healthy leaves 40-45 days after emergence, found 0.32–0.45% P in high-yielding plants compared with 0.10–

0.16% P in those giving low yields. Their values for deficient plants are very similar to that quoted for plants grown under glass by Roorda van Eysinga and Smilde (1969). Cheng and Forest (1977) quoted a diagnostic value of 0.38% P in mature cucumber leaves at the early blooming stage in the field. Jaszczolt and Nowosielski (1974) found the petioles of healthy bottom leaves to give the best guide to P status. The values reported were mainly for the inorganic P soluble in 2% acetic acid, corresponding to some 80% of the total petiole P in the July samples and 68% in September. The lower and upper critical levels for cucumber petioles were 0.15% and 0.68% inorganic P respectively, whilst values of up to 1.2% P were recorded, corresponding to decreased yields (cf. Massey et al., 1983).

It is evident from the references cited that published values for phosphorus in cucumber leaves show a very wide range. Some of the values quoted e.g. >1.0% are indeed so high as to suggest excessive accumulation of the element and possibly adverse effects on growth.

Sulphur

Deficiency symptoms and growth responses. Sulphur deficiency is virtually unknown in greenhouse cucumbers, since fertilisers such as potassium sulphate, magnesium sulphate and superphosphate supply more than enough of this nutrient for crop growth. Descriptions of S deficiency are, however, available based on experiments in sand and water culture.

Roorda van Eysinga and Smilde (1969) induced a pale green to yellow chlorosis in upper leaves where S was omitted. Growth was restricted and the leaves bent downwards. Ward (1976) studied responses to SO_4-S over the range 0–48 mg l^{-1} S, and also included an excessive level of SO_4-S (480 mg l^{-1}) in an attempt to induce toxicity. After 45 days, omission of S had decreased the height of the plants by 68%. The leaves of plants grown without S, were hard, very pale green-yellow, and curved downwards almost to the stem; the upper leaves were deeply serrated.

Holcomb and Hickman (1979) observed little stunting of S deficient plants, but marginal and interveinal areas of the leaf were chlorotic in contrast to the dark green of the main veins and adjacent tissue.

Sulphur content of the leaves. Roorda van Eysinga and Smilde (1969) found 0.06% S in S deficient leaves and 0.6–0.7% S in normal, healthy tissue. Ward (1976) showed a progressive increase in leaf content from 0.06% S where it was omitted to 0.6% at 48 mg l^{-1} SO_4–S in solution, the latter concentration being considered fully adequate: tissue levels below 0.25% S corresponded to severe deficiency.

Excess SO_4–S (480 mg l^{-1} S) caused downward curling of the leaf tips and some necrotic spotting, but the main response was depression of growth. Somewhat surprising-

ly, the S content of the upper leaves of young (45-day old) plants was not increased by excess SO_4-S compared with that found at the normal level of 48 mg l^{-1} S in solution. Even in older plants (77 days) the leaves from the 'high sulphate' treatment only averaged 1.41% S compared with 0.82% S in plants receiving the standard treatment, both these analyses being the average for all leaves on the main stem.

Potassium

Deficiency symptoms. The leaves of cucumber plants affected by K deficiency showed chlorosis, bronzing and marginal scorch (Purvis and Carolus, 1964). Malformation of the fruit was also described and illustrated, based on the work of I.C. Hoffman at the Ohio Agricultural Experiment Station; the fruits had enlarged tips but were undeveloped at the stem end. Roorda van Eysinga and Smilde (1969) described marginal chlorosis and bronzing, the chlorosis spreading interveinally. The leaf margins dried out and turned brown, but the veins remained green for some time. Growth was stunted and the leaves were small. Wetzold (1972) showed marginal browning which spread across the leaves as the deficiency became more severe, confining the green tissue to an area around the base of the main veins. Holcomb and Hickman (1979) reported that the leaves of deficient cucumber plants were pale green and smaller than normal. Tip burn was common, followed by bronzing and scorching of the leaves.

Growth responses. Potassium deficiency depressed the final yield of fruit by 45% (Adams et al., 1975a) and 51% (Adams et al., 1976). The deficiency reduced the early yield as the response on the main stem was greater than on the laterals (Adams et al., 1976); this supports a previous finding that correlations between yield and the K content of the leaves were highest during the early season and decreased as the season progressed (Bradley and Fleming, 1960). The grading of the fruit was also seriously affected by this deficiency; the fruit were reduced in size and the proportion of them acceptable in Class I decreased by 50% (Adams et al., 1976). Potassium deficiency caused a threefold increase in the proportion of poorly developed fruit (Adams, 1978). At the other extreme, very high levels also depressed yield (Adams et al., 1974a).

The uptake of K by plants in flowing solution culture (NFT) was closely related to light intensity (Tables 3 and 4) and to air temperature (Table 4), and was highly correlated with water usage (r = 0.96; Adams, 1980).

The K content of young cucumber shoots (5 weeks from sowing) was depressed progressively by increasing proportions of NH_4-N in the N supply (Barker and Maynard, 1972). The values ranged from 6.3% for plants receiving 100% NO_3-N to 3.5% with 100% NH_4-N, sampled after only two weeks of differential treatment in sand culture. Schenk and Wehrmann (1979b) showed a decrease in the

K content of young cucumber roots, though not of the shoots, with increasing concentrations of NH_4-N (0, 5 and 10 mM). Increasing concentrations of NH_3-N, achieved with various combinations of NH_4-N and pH, decreased the K content of both roots and shoots. NH_4-N greatly decreased the rates of K uptake and, at relatively high concentrations (0.09–0.28 mM) caused an efflux of K from the roots.

Potassium content of the leaves. Geissler (1957) reported 3.6% K in young cucumber leaves compared with 2.5% K in the older leaves. From nutritional trials with container-grown and field-grown pickling cucumbers Bradley *et al.* (1961) associated 3.1–3.7% K in the young expanded leaves with good growth and 1.8–2.5% K with low yields; the corresponding values for the oldest healthy leaves were 2.7–3.5% K and 1.4–1.7% K respectively. Lepiksaar (1965) reported 4.0–4.7% K in young, healthy plants. Ward (1967), studying the distribution of K within the plant, found an overall average of 4.44% K in the leaf laminae. The leaf analyses showed no clear positional trend, the values ranging from 4.0% to 5.5% K on the main stem and 4.1% to 5.4% K on the lateral shoots. Roorda van Eysinga and Smilde (1969) reported 2.5–5.4% K in the leaves of healthy plants. From nutritional experiments at Naaldwijk, Roorda van Eysinga (1970) found 3.47–4.33% K in cucumber leaves. The values reported by Wetzold (1972) for plants grown in sand culture were far higher in the young leaves (5.5–7.0% K) than in the older foliage (3.5–5.0% K). Ward (1973a) suggested a 'reference' value of 4.0% K in young cucumber laminae, and a 'diagnostic' value of 3.5% K was proposed for field-grown crops by Cheng and Forest (1977). Sonneveld gave a 'guide value' of 2.5% K in young, fully developed cucumber leaves.

The petioles of cucumber leaves contain far higher levels of K than do the laminae. Thus Ward (1967) found 8.5–14.8% K in the petioles from leaf positions 1–25 on the main stem, the lower values tending to occur in the lower part of the plant. Similarly, Jarrell and Johnson (1981) reported 15.3–16.6% K in the petioles as compared with 4.5–4.8% in the laminae.

Deficiency symptoms have been associated with levels of 0.5% K (Roorda van Eysinga and Smilde, 1969), 0.8–1.5% K (Wetzold, 1972) and of less than 2% K (Adams, 1982) in the leaves. Lepiksaar (1965) found 1.5% K in young cucumber plants affected by a deficiency. A reduction in yield was recorded, although there were no visible symptoms of deficiency on the foliage, when the leaves contained 2.1% K (Roorda van Eysinga and Smilde, 1969). A similar critical value of 2.3% was found by Adams (1982). The corresponding values for the contents of the petioles are 8.5% K or less for deficiency, 9% K for the critical level and 10–15% K for healthy leaves.

The K content of the sap pressed from freshly sampled

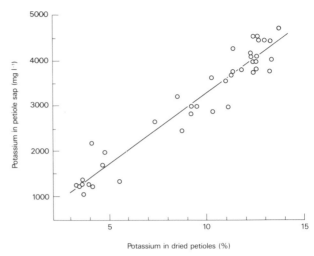

Figure 2 Relation between the potassium contents of dried cucumber petioles and of the sap pressed from petioles immediately after sampling (after Adams, 1982).

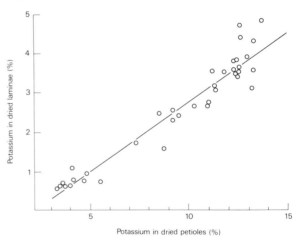

Figure 3 Relation between the potassium contents of the dried laminae and petioles of cucumber leaves (after Adams, 1982).

petioles was highly correlated with the K contents of the dried petioles (r = 0.96) and laminae (r = 0.97); see Adams (1982). Analysis of the petiole sap gave a rapid and accurate assessment of the K status of mature cucumber plants. Thus sap from the petioles of plants receiving adequate K contained 3500–5000 mg l^{-1} K. Loss of yield occurred below about 3000 mg l^{-1} K. Some relationships between the K contents of the laminae, petioles and petiole sap are shown in Figures 2 and 3.

Calcium

Whilst Ca deficiency is not a common disorder of cucumbers, its occurrence amongst glasshouse crops in Finland was noted by Hårdh (1957). Roorda van Eysinga and Smilde (1969) referred to a deficiency being associated with soils depleted of Ca by leaching, and also with the use of sphagnum peat as a rooting medium. Calcium moves in the xylem but is present only at low concentrations in the phloem sap. There is thus little translocation from the older to the younger leaves. When the supply of this nutrient is interrupted or becomes depleted, deficiency symptoms are thus usually seen at the very top of the plant.

Deficiency symptoms and growth responses. The symptoms of Ca deficiency in cucumbers are well documented. Hoffman (1933), quoted by McMurtrey (1948), described mottled yellowing and brown spotting of the leaves; the plants were stiff and stunted. Hårdh (1957) gave a comprehensive account embodying previously published descriptions. The roots of Ca deficient cucumber plants were poorly developed, thicker and shorter than normal, turned brown at an early stage and had few root hairs. The margin of the upper leaves became pale green and turned upwards. The interveinal areas became yellow and then necrotic. The leaves were small and the stems thin with few secondary shoots. The flowers were small and pale, and the fruits small and tasteless. Illustrations by Roorda van Eysinga and Smilde (1969) show white spots near the veins of the leaves and interveinal yellowing which spread as the deficiency became more severe. They also noted that the internodes became very short and the new leaves remained small, the upcurled margins of which gradually dried out. In contrast, the margins of the older leaves were downcurled. The flower buds abort when the deficiency is severe and the growing point dies. Wetzold (1972) shows downcurling of the margins of all but the smallest leaves, a greyish-brown scorch in the head of the plant and failure of the fruit to develop normally at the blossom end. Ward (1973b) described interveinal chlorosis and chlorotic spotting, accompanied by downward curling and hooking of the leaves. The plants were stunted and the growing points aborted. Fruit size decreased with decreasing Ca concentration in sand culture over the range 0–80 mg l^{-1} Ca, whereas the numbers of fruit set were affected only under conditions of severe deficiency (0–8 mg l^{-1} Ca). Severe downward cupping and distortion of the leaves were recorded by Holcomb and Hickman (1979), together with a marginal chlorosis which sometimes extended up between the main veins. Very severe deficiency symptoms were described by Matsumoto and Teraoka (1980), working in aerated solution culture. When Ca was omitted from the nutrient solution at the 4–5 leaf stage, growth was almost completely inhibited, the roots had a dull appearance and the tops withered within 4–6 days. The margins of the young leaves curled upwards and inwards. The content of soluble sugars increased greatly (up to ten-fold) in Ca deficient leaves but decreased in the roots.

Calcium content of the leaves. The Ca content of the leaves was depressed by over 40% when only NH_4-N was supplied (Barker and Maynard, 1972), and was increased

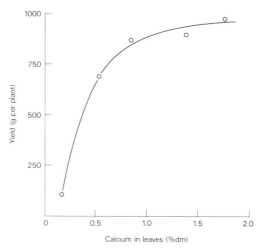

Figure 4 Relation between the calcium content of the leaves and the yield of cucumber fruit (cv. Sporu; after Ward, 1973).

slightly (12%) when peat was limed from pH 5.2 to pH 6.7 (Prasad and Byrne, 1975). The relationship between the yield of fruit obtained 67 days after sowing and the Ca content of the top leaf, based on Ward's data (1973b), is shown in Figure 4.

Ward (1967) found that the Ca content of the leaves increased from 1.1% Ca in the top leaf to 13.1% Ca in the bottom leaves; the corresponding values for the petioles were 1.1% Ca and 6.5–7.9% Ca. Severe deficiency was associated with 0.2% Ca or less in the top leaves (Ward, 1973b), though it was difficult to define a clear threshold value because the symptoms developed gradually. Somewhat higher values associated with the deficiency are suggested elsewhere. For example, Wetzold (1972) found deficiency below 0.7% Ca in young leaves and below 2.0% Ca in old leaves. Roorda van Eysinga and Smilde (1969) recorded 2.3% Ca in leaves with symptoms and 4.3% or less in leaves from plants giving a reduced yield but having no foliar symptoms. It was suggested that leaves from healthy plants should contain 5.7–11.4% Ca, the same authors (1981) later quoting a slightly lower range of 2.0–10.0% Ca. Ward (1973a) proposed a reference standard of 1.5% Ca in the young leaves (third below the top of the plants), and Sonneveld (1981) suggested a 'guide value' of 5.0% Ca in the young, fully developed leaves.

Magnesium

Symptoms of deficiency and excess. Magnesium deficiency is a common disorder of cucumbers, both in the greenhouse and in the field. Thus Hoffman (1933), quoted by McMurtrey (1948), described a mottled chlorosis and brown spotting of the leaves. Carolus (1934) referred to interveinal yellowing with the veins remaining green, usually occurring first on the lower leaves; the leaf edges were described as ragged. In many cases (see Roorda van Eysinga and Smilde, 1969, and Bergmann, 1976) a green margin remains around the leaf even when the deficiency is severe and parts of the yellow interveinal areas have dried out to a pale brown colour. This is not always the case, however, and Wetzold (1972) showed some leaves with green areas around the main veins only, the rest, including the margins, being yellow. The persistence of a green margin around the leaf in some, though by no means all, instances of magnesium deficiency in cucumbers finds its parallel with tomatoes (see p. 60).

Toxicity symptoms, consisting of marginal scorching on dark green leaves, were reported at 896 mg l^{-1} Mg in solution culture (Hanada *et al.*, 1981).

Growth responses. In a factorial trial, Mg deficiency reduced the overall yield by 9% (Adams *et al.*, 1975a) and 17% (Adams *et al.*, 1976). However, interactions with other nutrients resulted in greater depressions in yield. For example, the yield from deficient plants declined progressively as the level of applied N increased (see Table 5) and was also reduced by up to 20% at high levels of applied K. Magnesium deficiency nearly doubled the proportion of poorly developed fruit (Adams, 1978). Hanada *et al.* (1981) and Kim *et al.* (1981) found a decreased rate of photosynthesis in Mg deficient cucumber leaves. The effect was detectable at a concentration of 12.2 mg l^{-1} Mg in solution culture and very marked at 0 and 6.1 mg l^{-1} Mg.

Nitrogen content of liquid feed (mg l^{-1})	Magnesium level		Loss in yield due to magnesium deficiency (%)
	Added	Omitted	
50	13.5	12.9	4.4
175	15.9	13.0	18.2
300	16.5	11.9	27.9

Table 5 Effects of nitrogen and magnesium on the yield of cucumbers. (after Adams, 1978).

Magnesium content of the leaves. Magnesium, being translocated in the phloem, is relatively mobile within the plant. Carolus (1934) showed that Mg deficient plants had a higher content of this element in their upper leaves (0.22% Mg) than in the lower leaves (0.13% Mg). Where it was supplied to the crop, however, it accumulated particularly in the lower leaves (0.77% Mg) compared with the upper foliage (0.46% Mg). Geissler (1957) reported 0.5% Mg in young leaves and 0.9% Mg in the older foliage. In a detailed study of the Mg distribution in cucumber plants adequately supplied with this nutrient, Ward (1967) found 0.6% Mg in the youngest leaves compared with up to 1.5–2.0% Mg in the very oldest leaves. The average for all leaves, including both those on the main stems and on the laterials, was 0.76% Mg. Roorda van Eysinga and Smilde (1969) found 0.22% Mg in deficient leaves compared with 0.6–1.3% Mg in healthy leaves. Wetzold (1972) gave values of 0.5–0.9% Mg in young leaves from healthy plants, with deficiency below 0.35% Mg. The corresponding values for the older leaves

were 0.8–1.8% Mg, with deficiency below 0.4% Mg. Barker and Maynard (1972) showed a marked depression in the Mg content of young cucumber shoots by NH_4-N, the values ranging from 0.67% Mg when all the N was supplied as NO_3-N to 0.44% Mg with 100% NH_4-N; an intermediate value of 0.55% Mg was found when both forms were present. Ward (1973a) proposed a 'reference' standard of 0.45% Mg in the young leaf laminae (3rd below top), whilst Sonneveld (1981) gave a 'guide value' of 0.8% Mg.

Iron

The youngest leaves of plants sown without Fe were bright yellow-green, later becoming almost ivory in colour, and the plants remained in a state of suspended development (Hewitt and Watson, 1980). A fuller description was given by Roorda van Eysinga and Smilde (1969): the youngest leaves first became yellow-green or yellow but the veins remained green. As the severity of the deficiency increased the veins gradually became yellow until the whole leaf was lemon yellow or yellowish-white. Growth ceased, the leaf margins became brown and the fruit became pale in colour (see Roorda van Eysinga and Smilde, 1981). Young plants with yellow leaves but green veins were illustrated by Wetzold (1972). Holcomb and Hickman (1979) described a prominent interveinal chlorosis of the leaves, most severe at the margins. Leaves formed subsequently showed progressively more severe symptoms, resulting in a general yellowing.

Applications of different levels of Fe to sphagnum peat had little effect on the yield of fruit (Sonneveld and Voogt, 1975). A similar lack of response by cucumbers grown in peat has been noted by the present authors, though the yield was slightly reduced on one occasion by omission of Fe (Adams *et al.*, 1974).

Iron content of the leaves. Iron was reported to accumulate in the lower leaves (Nollendorfs and Upits, 1972). These authors suggested that deficiency symptoms were likely to occur when the 4th or 5th leaf from the top contained less than 100–120 μg g^{-1} Fe; healthy growth and high yields of fruit were associated with 150–250 μg g^{-1} Fe. Wetzold (1972), working with plants grown in sand culture, found 100–300 μg g^{-1} Fe in healthy leaves, with deficiency below 50 μg g^{-1} Fe. Sonneveld (1981) suggested that young, fully developed cucumber leaves should contain 100 μg g^{-1} Fe, though chlorosis might still occur at higher values.

As noted for tomatoes, the total Fe content is not a reliable assessment of the amount of Fe that is physiologically active within the leaves, and an enzymic test may be more suitable. Besford (1975) showed that the activity of peroxidase was reduced by 74% in plants not supplied with Fe. However, the interpretation of this test is complicated by the fact that the activity of peroxidase is increased threefold by boron deficiency.

Manganese

Deficiency symptoms and growth responses. Skinner (1941) described interveinal chlorosis of the leaves, the veins remaining green. Growth was weak, and the flower buds turned yellow. The symptoms were described in greater detail by Roorda van Eysinga and Smilde (1969). At first even the smallest veins remain green, producing a fine reticular green pattern on a yellow background. Later, with the exception of the main veins, the entire leaf becomes yellow and whitish sunken areas develop between the veins. Further illustrations of most of these symptoms have been given by Wetzold (1972) and by Bergmann (1976). Holcomb and Hickman (1979) recorded that their Mn deficient plants, grown in aerated solution culture, had a general pale green colour, with slender stems and shortened internodes. Recently matured leaves showed a network of green veins with dull yellow-green interveinal tissue, though the symptoms were not particularly striking. Cucumber was the most susceptible of several species when raised from seed in the absence of Mn (Hewitt and Watson, 1980); chlorosis and twisting of the leaves was followed by death within three weeks. When transferred from complete nutrient solution to one containing no Mn, faint general mottling developed in the rapidly expanding leaves within days.

Manganese deficiency of cucumbers occurs with some calcareous soils and on heavily limed sand or peat soils (Roorda van Eysinga and Smilde, 1969). Omission of Mn from a peat substrate reduced cucumber yields by 6% (Sonneveld and Voogt, 1975). The highest yields were obtained with 25 g m^{-3} Mn applied before planting and a further 18 g m^{-3} in the top dressings.

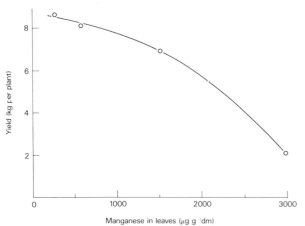

Figure 5 Relation between the manganese content of the leaves and the yield of cucumber fruit obtained in sand culture (after Messing; unpublished data).

Manganese content of deficient and healthy leaves. The Mn content of cucumber leaves associated with maximum yield was given by Sonneveld and Voogt (1975) as 350 μg g^{-1}, this value being considerably above other published data. Thus yields were not reduced when the leaves contained 67–251 μg g^{-1} Mn (Adams; unpublished data) and good yields have also been associated with 40–120 μg g^{-1} Mn (Nollendorfs and Upits, 1972). Roorda van Eysinga and Smilde (1969) reported 100–300 μg g^{-1} Mn in healthy leaves. A recommended level of 60–150 μg g^{-1} Mn in the leaves was quoted from the Saanichton Research Station in Canada (Anon., 1981), and Sonneveld (1981) quoted a 'guide value' of 50 μg g^{-1}. Manganese deficient leaves were found to contain 12 μg g^{-1} Mn (Roorda van Eysinga and Smilde, 1969), and a similar value was reported by Bergman (1976). Wetzold (1972) suggested that deficiency occurred below 15 μg g^{-1} Mn in the young leaves and 20 μg g^{-1} in the old leaves. The Mn content of the leaves is generally higher towards the bottom of the plant. Thus Wetzold (1972) found 30–60 μg g^{-1} Mn in young leaves compared with 100–250 μg g^{-1} in the older foliage; see also the regressions of Mn content of young and old leaves on the content of this nutrient in the substrate, published by Sonneveld and Voogt (1975).

Manganese toxicity. The symptoms of Mn toxicity first appear on the oldest leaves. Pale green and yellow areas develop between the veins and dry out as the condition becomes more severe. The veins become red/brown in colour, and numerous purple spots develop on the stems and petioles (Roorda van Eysinga and Smilde, 1969). Wetzold (1972) also referred to spotting of the petioles and of the veins on the underside of the leaves, and noted a violet colour in the accessory cells of the plant hairs. Yellowing of the laminae developed around the smaller veins rather than in between them.

Studies in sand culture (Messing, 1964) showed that growth was reduced and characteristic symptoms of toxicity developed when the young leaves contained more than about 500 μg g^{-1} Mn. The relationship between yields of fruit and the Mn content of the leaves is shown in Figure 5. Roorda van Eysinga and Smilde (1969) reported toxicity at or above 682 μg g^{-1} Mn in the leaves. The corresponding values obtained by Wetzold (1972) were 500 μg g^{-1} in young leaves and 800 μg g^{-1} in the older foliage. Depressions in yield of up to 32% were recorded by Sonneveld and Voogt (1975) where excessive amounts of manganese sulphate had been applied to a peat substrate; leaf analysis showed about 2000–5000 μg g^{-1} Mn in the young and old leaves respectively.

Manganese toxicity is frequently associated with partial steam sterilisation of soil, particularly if the latter is acidic or the steaming is unduly prolonged; see Messing (1971). The response to steaming temperature was studied by Messing (1966) over the range 65–100°C, using steam-air mixtures. The Mn content of the leaves increased progressively from 105 μg g^{-1} in unsteamed soil to 806 μg g^{-1} where the soil had been steamed at 100°C. The highest yields were obtained in soil steamed at 77–88°C, and at 100°C the yield was depressed by 8%.

Copper

Deficiency symptoms. Deficiency of Cu restricted the growth of cucumbers, giving short internodes and small leaves. Interveinal chlorotic blotches appeared on the older leaves, the symptoms spreading upwards on the plant. The leaves eventually turn dull green or bronze colour and wither (Roorda van Eysinga and Smilde, 1969). Nollendorfs and Upits (1972) observed very acute Cu deficiency in a peat substrate. Neither buds nor flowers developed at the tops of the plants. The edges of the leaves turned downward and the plants were dwarfed. Adams *et al.* (1978b) described a yellow lace-like pattern which developed near the margin of the leaves of deficient plants soon after planting. The yellowing gradually spread across the leaves, leaving only small, isolated areas of green tissue; for illustration see Adams (1978). The interveinal areas of the lower leaves became severely scorched and the leaves collapsed and withered. Holcomb and Hickman (1979) confirmed the stunting and reduction in leaf size associated with Cu deficiency, and described small white necrotic areas on the leaves.

Growth responses. Four weeks after planting, Cu deficiency had reduced the height of plants grown in peat by 32% (Adams *et al.*, 1978b). The final yields were greatly reduced e.g. by 94% (Adams *et al.*, 1974b), by 50% (Adams *et al.*, 1975b), by 88% (Adams *et al.*, 1976), and, for four crops, by an average of 48% (Adams *et al.*, 1978b). The early yields were also much reduced e.g. by 74% (Adams *et al.*, 1978b), 90% and 95% (Adams *et al.*, 1978a) 93% (Graves *et al.*, 1979). Over 88% of the fruits set on deficient plants failed to develop properly; 58% of the fruit were poor in colour (Graves *et al.*, 1979) and the proportion of poorly shaped fruit increased four-fold (Adams *et al.*, 1978a). The average weight per fruit set was reduced by 14% (Adams *et al.*, 1978b).

Although this deficiency is particularly associated with peat substrates, responses to the application of copper sulphate have also been found on a fine sandy soil containing 3.2% organic matter (Navarro and Locascio, 1973). In the first year, the early yield of field-grown cucumbers increased markedly with the level applied and a four-fold increase in the final yield was obtained at the highest rate of application. In the following year the yields from all treatments were much higher, but there was still a yield depression of 45% where Cu was not supplied.

Copper content of the leaves. The highest yields, averaged for four crops of cucumbers, were associated with about 9 μg g^{-1} Cu in the fifth leaf from the tops (Adams *et al.*, 1978b). Normal leaves contained 7–10 μg g^{-1} Cu (Roorda van Eysinga and Smilde, 1969), 10–18 μg g^{-1} (Nollendorfs and Upits, 1972) and 8–20 μg g^{-1} (Adams, 1978). Sonneveld (1981) proposed a 'guide value' of 7 μg g^{-1} Cu. Copper deficient leaves were found to contain 2

μg g^{-1} Cu (Roorda van Eysinga and Smilde, 1969), and values below 6.4 μg g^{-1} were regarded as indicating deficiency (Roorda van Eysinga and Smilde, 1981). Copper contents of 6–8 μg g^{-1} in the 4–5th leaf from the top were classed as 'insufficient' by Nollendorfs and Upits (1972), and a far lower value of 1.8 μg g^{-1} Cu was found in leaves with acute deficiency symptoms. Adams *et al.* (1978b) found an average yield reduction of 23% for plants containing 1.9 μg g^{-1} Cu in their leaves, with values of 0.8–1.2 μg g^{-1} in cases of severe deficiency.

Zinc

Symptoms and growth responses. Zinc deficiency is a rare disorder and the symptoms have not been fully described. A slight interveinal mottle developed on the lower leaves and spread up the plant, and the upper internodes failed to extend (Roorda van Eysinga and Smilde, 1969). Holcomb and Hickman (1969) also found shortening of the internodes and decreased leaf size but with no distinctive leaf colour. Hewitt and Watson (1980) noted that the leaves were generally pale green and that only the main veins and a narrow band of tissue beside them retained their normal colour.

In peat, Sonneveld and Voogt (1975) found that moderate applications of Zn increased the yield by up to 32%, whereas heavy dressings depressed it by 17%. The maximum yield was associated with 200 μg g^{-1} Zn in the leaves. In another instance, however, omission of Zn from the peat had little effect on yield (Adams *et al.*, 1974).

Zinc content of the leaves. Healthy growth is associated with 40–80 μg g^{-1} Zn (Nollendorfs and Upits, 1972), 40–100 μg g^{-1} (Adams, 1978) and 50 μg g^{-1} (Sonneveld, 1981) in the leaves. Symptoms of Zn deficiency occur when the fifth leaf contains less than 20–25 μg g^{-1} Zn (Nollendorfs and Upits, 1972). Roorda van Eysinga and Smilde (1969) found only 9 μg g^{-1} Zn in the leaves of deficient plants grown in water culture.

Excessive levels induce symptoms similar to those of Fe deficiency, namely a yellowing of the leaves in the head of the plant which gradually spreads to the older leaves (see Roorda van Eysinga and Smilde, 1969); these authors found 900 μg g^{-1} Zn in the tops of affected plants. Nollendorfs and Upits (1972) reported toxicity symptoms at levels above 150–180 μg g^{-1} Zn in the oldest leaves.

Boron

Deficiency symptoms. Mahrotra *et al.* (1969), working with plants grown in sand and water culture, reported a dark green colour and leathery texture in the leaves, followed by death of the apical bud. The older leaves later developed a brownish-yellow interveinal chlorosis and scorched margins, whilst the younger leaves were often malformed and cupped, with prominent veins. The development of axillary buds resulted in a squat, bushy appearance. Roorda van Eysinga and Smilde (1969) recorded death of the very young unexpanded leaves and of the growing point; the older leaves became stiff and cupped upwards. Generally similar features were illustrated by Wetzold (1972), but without cupping of the older leaves of a mature plant. Holcomb and Hickman (1979) observed various forms of leaf distortion, including puckering of internal leaf surfaces, fluted margins and downcurling. The leaves were brittle and reduced in size. After the growing point has died, the mature leaves become enlarged and are both coarse and brittle (Adams and Winsor, unpublished data). Fruits from deficient plants may be rather short (see Wetzold, 1972), and frequently develop longitudinal corky cracks in the skin similar to those caused by low air temperatures (Adams, 1978).

Growth responses. Prasad and Byrne (1975) studied the yields from young cucumber plants grown in peat at four levels of B in combination with three levels of liming. Maximum yields were obtained from the pot-grown plants when B was added at 3.2 g m^{-3}. Omission of B decreased the yields by 20%, 94% and 95% at pH 5.2, 6.3 and 6.7 respectively, showing a marked interaction with liming. Plants grown in sphagnum peat without added B were stunted, though growth was continued by lateral shoots, and the yield was depressed by 70% (Adams *et al.*, 1974b). Boron deficiency delayed fruit formation and reduced the weight per fruit by 10% and the final yield by 31% (Adams *et al.*, 1976). When the deficiency was less severe and reduced the final yield by only 8%, it nevertheless caused a marked reduction (39%) in the yield during the first four weeks of harvesting, and increased the number of poorly developed fruit by 150% (Graves *et al.*, 1979). In other instances, however, added B had little effect on yield (Sonneveld and Voogt, 1975; Adams *et al.*, 1977).

Boron content of the plants. The distribution of B in cucumber plants was studied by Besford and Manning (1981) using solution culture. They found that the concentrations in the dry matter of the apex, leaves, stem and root all declined after B was withheld whereas the total B content of these tissues remained almost constant except in the apex, where it declined progressively. The data supported other work showing that B is not re-translocated within the plant and that a continuous supply of this nutrient to the roots is essential for healthy growth of new tissues.

Boron-deficient leaves were found to contain 20 μg g^{-1} B (Roorda van Eysinga and Smilde, 1969), less than 20–25 μg g^{-1} (Nollendorfs and Upits, 1972), 25 μg g^{-1} (Wetzold, 1972) and 19 μg g^{-1} B (Manning and Besford, 1978). Healthy leaves contain 40–120 μg g^{-1} B (Roorda van Eysinga and Smilde, 1969), 30–60 μ g g^{-1} (Nollendorfs

and Upits, 1972), 50–140 μg g^{-1} (Wetzold, 1972), 30–110 μg g^{-1} (Sonneveld and Voogt, 1975), 40–80 μg g^{-1} (Adams, 1978), 56–80 μg g^{-1} (Manning and Besford, 1978) and 50 μg g^{-1} B (Sonneveld, 1981).

Boron toxicity. Cucumber plants are very sensitive to high levels of B in the substrate or water supply. Growth was reduced by high levels in the peat (Prasad and Byrne, 1975) and the dry weight of the plants was depressed by over 50% when the culture solution contained excessive concentrations of B (El-Sheikh *et al.*, 1971).

With B toxicity, the margins of the older leaves become yellow (see Bergmann, 1976), curl downwards and gradually become brown (see Roorda van Eysinga and Smilde, 1969). The leaves from plants with symptoms contained 300–500 μg g^{-1} B (Roorda van Eysinga and Smilde, 1969) and 240–280 μg g^{-1} (Nollendorfs and Upits, 1972). Growth was much reduced when the laminae of mature leaves contained more than 400 μg g^{-1} B (El-Sheikh *et al.*, 1971). Wetzold (1972) recorded toxicity at above 400 μg g^{-1} B in the young leaves and 800–1000 μg g^{-1} in the old leaves. Symptoms of toxicity were not observed when the content of the young leaves was 110 μg g^{-1} B and that of the older ones was 250 μg g^{-1} (Sonneveld and Voogt, 1975). Prasad and Byrne (1975) found that toxicity symptoms developed when the plants (all parts combined) contained 115 μg g^{-1} B. The young leaves contained only 50 μg g^{-1} B, however, when mild symptoms of toxicity developed in the older leaves (Manning and Besford, 1978).

Molybdenum

Deficiency symptoms. Johnson (1966), quoting the work of von Stieglitz and Chippendale (1955), described yellow patches on the older leaves of Mo deficient plants; the leaves became scorched, with uprolled margins. Blake (1959) noted stunting of deficient plants; the leaves showed marginal and interveinal chlorosis (pale or yellow-green) followed by marginal scorch. Roorda van Eysinga and Smilde (1969) described the development of a pale green colour in the leaves of deficient plants, particularly between the veins. The leaves eventually became yellow and died, the symptoms beginning on the lower leaves and spreading up the plant. Growth was hardly affected but the flowers were small. Holcomb and Hickman (1979) noted death of the older leaves without the marginal scorching or leaf rolling often associated with Mo deficiency in other plants. Leaf size was reduced and the plants were spindly with shortened internodes. According to Hewitt and Watson (1980) the basal areas of cucumber leaves first become chlorotic, with an irregular green zone between the central and chlorotic outer regions and often a bright green, narrow marginal rim; the leaves later become bleached and scorched.

Growth responses. Plants grown in sphagnum peat without added Mo were stunted, with pale green leaves. Small yellow areas developed on the upper leaves and gradually spread across the interveinal areas; in acute cases these areas were pale cream or white and the lower leaves, which tended to wilt, gradually collapsed and withered. The deficiency reduced the final yield of fruit by 84% (Adams *et al.*, 1976). In the following year, this deficiency reduced plant height by 15% and the average dry weight per leaf by 19% (56% for severely affected leaves). The early yield was depressed by 60% and the dry matter content of the fruit by 19% (Adams *et al.*, 1977). The yield of cucumbers was increased by 99% when the pH of a peat substrate, not supplied with Mo, was raised from 5.1 to 6.7 by liming (Adams, 1985).

Molybdenum content of the leaves. Deficient leaves contain less than 0.3 μg g^{-1} Mo (Roorda van Eysinga, 1969), 0.5–0.6 μg g^{-1} (Nollendorfs and Upits, 1972) and 0.5 μg g^{-1} (Adams, 1978), and have rather higher NO_3-N contents than healthy leaves (Adams, 1985). The leaves from healthy plants contain 0.8–3.3 μg g^{-1} Mo (Roorda van Eysinga and Smilde, 1969) and 1–5 μg g^{-1} Mo (Nollendorfs and Upits, 1972).

References

Adams, P. (1978). How cucumbers respond to variation in nutrition. *Grower* **89**, 197, 199, 201.

Adams, P. (1980). Nutrient uptake by cucumbers from recirculating solutions. *Acta Hort* **98**, 119–126.

Adams, P. (1982). Assessing the potassium status of cucumber plants. *Proc 9th Int Pl Nutrit Colloq Coventry 1982: Commonw Agric Bur*, 7–11.

Adams, P. (1985). Some responses of cucumbers, grown in beds of peat, to copper and molybdenum. *Acta Hort* **156**, 73–80.

Adams, P. and Winsor, G. W. (1984). Some responses of cucumbers and lettuce, grown in a peat substrate, to phosphate and lime. *Acta Hort* **145**, 102–111.

Adams, P., Graves, C. J. and Winsor, G. W. (1974a). Nutrition of cucumbers in peat: Responses to nitrogen, potassium and magnesium. *Rep Glasshouse Crops Res Inst 1973*, 71.

Adams, P., Adatia, M. H., Graves, C. J. and Winsor, G. W. (1974b). Nutrition of cucumbers in peat: Responses to individual micronutrients. *Rep Glasshouse Crops Res Inst 1973*, 72.

Adams, P., Graves, C. J. and Winsor, G. W. (1975a). Nutrition of cucumbers in peat: Responses to nitrogen, potassium and magnesium. *Rep Glasshouse Crops Res Inst 1974*, 70.

Adams, P., Adatia, M. H., Graves, C. J. and Winsor, G. W. (1975b). Nutrition of cucumbers in peat: Responses to individual micronutrients. *Rep Glasshouse Crops Res Inst 1974*, 70.

Adams, P., Graves, C. J. and Winsor, G. W. (1976). Nutrition of cucumbers in peat. *Rep Glasshouse Crops Res Inst 1975*, 70–72.

Adams, P., Graves, C. J. and Winsor, G. W. (1977). Nutrition of cucumbers in peat. *Rep Glasshouse Crops Res Inst 1976*, 78–79.

Adams, P., Graves, C. J. and Winsor, G. W. (1978a). Nutrition of cucumbers in peat. *Rep Glasshouse Crops Res Inst 1977*, 80–81.

Adams, P., Graves, C. J. and Winsor, G. W. (1978b). Effects of copper deficiency and liming on the yield, quality and copper status of tomatoes, lettuce and cucumbers grown in peat. *Scientia Hort* **9**, 199–205.

Anon (1981). Micronutrient concentration for tomato and long English cucumbers. *Saanichton Res Plant Quar Stn, Hi-lites 1980*, 8–9.

Barker, A. V. and Maynard, D. N. (1972). Cation and nitrate accumulation in pea and cucumber plants as influenced by nitrogen nutrition. *J Am Soc hort Sci* **97**, 27–30.

Bergmann, W. (1976). Ernährungsstörungen bei Kulturpflanzen in Farbbildern. *VEB Gustav Fisher Verlag, Jena* 183 pp.

Bergmann, W., Büchel, L. and Wrazidlo, W. (1965). Bor-und Stickstoff-Uberschusssymptome bei Gewächshausgurken sowie sowie Borschäden bei einigen anderen Pflanzen. *Arch Gartenb* **13**, 65–76.

Besford, R. T. (1975). Enzymes as indicators of crop nutritional status. *Rep Glasshouse Crops Res Inst 1974*, 59–60.

Blake, C. D. (1959). Molybdenum deficiency in cucumbers. *Banana Bull* **22**, 7.

Bradley, G. A. and Fleming, J. W. (1960). The effects of position of leaf and time of sampling on the relationship of leaf phosphorus and potassium to yield of cucumbers, tomatoes and melons. *Proc Am Soc hort Sci* **75**, 617–624.

Bradley, G. A., Fleming, J. W. and Mayes, R. L. (1961). Yield and quality of pickling cucumbers and cantaloupes as affected by fertilisation. *Bul Ark agric Exp Stn* **643**, 23 pp.

Carolus, R. L. (1934). Effects of magnesium deficiency in the soil on the yield, appearance and composition of vegetable crops. *Proc Am Soc hort Sci* **32**, 610–614.

Cheng, B. T. and Forest, B. (1977). The nutrition of *Cucumis sativa* L. *Agrochim* **21**, 286–294.

Dearborn, R. B. (1936). Nitrogen nutrition and chemical composition in relation to growth and fruiting of the cucumber plant. *Mem Cornell Univ agric Exp Stn* **192**, 26 pp.

El-Sheikh, A. M. and Broyer, T. C. (1970). Concentrations of total nitrogen in squash, cucumber and melon in relation to growth and to a Piper-Steenbjerg effect. *Commun Soil Sci Pl Anal* **1**, 213–219.

El-Sheikh, A. M., El-Hakam, M. A. A. and Ulrich, A. (1970). Critical nitrate levels for squash, cucumber and melon plants. *Commun Soil Sci Pl Anal* **1**, 63–74.

El-Sheikh, A. M., Ulrich, A., Awad, S. K. and Mawardy, A. E. (1971). Boron tolerance of squash, melon, cucumber and corn. *J Am Soc hort Sci* **96**, 536–537.

Geissler, T. (1957). Der Nährstoffentzug einer frühen Treibgurkenkultur. *Arch Garten* **5**, 431–465.

Graves, C. J., Adams, P. and Winsor, G. W. (1979). Nutrition of cucumbers in peat. *Rep Glasshouse Crops Res Inst 1978*, 86–87.

Hanada, T., Takahashi, K., Nagaoka, M. and Yoshioka, H. (1981). Effects of nitrogen, magnesium and phosphorus concentration in nutrient solution on the photosynthesis of cucumber leaves. *Bull Veg Ornam Crops Res Stn* Series A No. 9, 83–96.

Hårdh, J. E. (1957). On the calcium uptake of glasshouse cucumber. *Maataloust Aikakausk* **29**, 238–242.

Hewitt, E. J. and Watson, E. F. (1980). The production of micronutrient element deficiencies in plants grown in recirculated nutrient film (NFT) systems. *Acta Hort* **98**, 179–189.

Hoffman, I. C. (1933). Mineral deficiency symptoms in tomato and cucumber plants. *Ohio Veg Gro Assoc Proc* **18**, 58–62.

Holcomb, J. S. and Hickman, G. W. (1979). Greenhouse cucumber nutrient deficiency determination. *Proc 1979 Ann Western Greenhouse Veg Gro Conf Fresno Calif*, 24–31.

Jarrell, W. M. and Johnson, Jr., H. (1981). Comparative mineral nutrition of greenhouse cucumbers grown in soil mix and flowing nutrient solution. *Proc 1981 Ann Western Greenhouse Veg Gro Conf Fresno Calif*, 37–43.

Jaszczolt, E. and Nowosielski, O. (1974). Investigations on indices of nourishing cabbage and cucumber with phosphorus. Part 1. Critical content of mineral phosphorus in denotative parts of cabbage and cucumber. *Roczn Nauk roln*, Seria A, **100**, 29–44.

Johnson, C. M. (1966). In H. D. Chapman, ed. Diagnostic Criteria for plants and soils, 286–301. *Univ Calif, Div agric Scis.*

Kim, K. Y., Hanada, T. and Takahashi, K. (1981). Effects of magnesium content in culture solution on the photosynthetic activity of cucumber leaves. *Res Reports, Off Rur Dev (Hortic and Seric)* **23**, 1–9.

Lepiksaar, J. (1965). Die Düngung von Gewächshausgurkenjungpflanzen bei Anzucht in Torfkultursubstrat. *Arch Gartenbau* **13**, 35–52.

McMurtrey, J. E. (1948). Visual symptoms of malnutrition in plants. In H. B. Kitchen, ed. Diagnostic Techniques for Soils and Crops, 231–285. *The American Pot Institute, Washington.*

Mahrotra, O. N., Saxena, H. K. and Misra, P. H. (1969). Studies on mineral nutrition of cucurbits. I. Boron deficiency in cucumber – *Cucumis sativus* L. variety Kalyanpur local. *Ind J Hort* **26**, 69–73.

Manning, K. and Besford, R. T. (1978). Effect of boron nutrition on some enzymes in the cucumber plant. *Rep Glasshouse Crops Res Inst 1977*, 66–67.

Manning, K. and Besford, R. T. (1981). Boron supply and uptake by cucumber plants grown in solution culture. *Rep Glasshouse Crops Res Inst 1980*, 53–55.

Massey, D. M. (1978). Response of cucumbers to nitrogen concentration. *Rep Glasshouse Crops Res Inst 1977*, 78.

Massey, D. M., Goodman, A. C. and Winsor, G. W. (1983). Effect of phosphorus concentration on cucumbers. *Rep Glasshouse Crops Res Inst 1981*, 46–47.

Matsumoto, H. and Teraoka, K. (1980). Accumulation of sugars in cucumber leaves during calcium starvation. *Pl Cell Physiol* **21**, 1505–1513.

Messing, J. H. L. (1964). Manganese toxicity. *Rep Glasshouse Crops Res Inst 1963*, 63–64.

Messing, J. H. L. (1966). Partial sterilisation of the soil with steam/air mixtures. *Rep Glasshouse Crops Res Inst 1965*, 77–78.

Messing, J. H. L. (1971). Metal toxicities in glasshouse crops. *Tech Bull Minist Agric Fish Fd* **21**, 159–175.

Navarro, A. A. and Locascio, S. J. (1973). Cucumber response to copper rate and fertiliser placement. *Proc Fla State Hort Soc* **86**, 193–195.

Nollendorfs, V. and Upits, V. (1972). Optimization of trace element content in sphagnum peat substrate for underglass cucumbers. *Proc 4th Int Peat Cong Otaneimi Finland*, 273–282.

Prasad, M. and Byrne, E. (1975). Boron source and lime effects on the yield of three crops grown in peat. *Agron J* **67**, 553–556.

Purvis, E. R. and Carolus, R. L. (1964). Nutrient deficiencies in Vegetable crops. *In* H. B. Sprague, ed. Hunger Signs in Crops, 3rd Ed. 245–286. *David McKay Co N.Y.*

Roorda van Eysinga, J. P. N. L. (1970). Bemestingsproeven met konkommer in de periode 1964–1969. *Instit Bodensvruchtb, Haren-Groningen Rapport* **13**, 22pp.

Roorda van Eysinga, J. P. N. L. and Smilde, K. W. (1969). Nutritional disorders in cucumbers and gherkins under glass. *Cent agric Publ and Documn, Wageningen*, 46 pp.

Roorda van Eysinga, J. P. N. L. and Smilde, K. W. (1981). Nutritional disorders in glasshouse tomatoes, cucumbers and lettuce. *Cent agric Publ and Documn, Wageningen*, 130 pp.

Schenk, M. and Wehrmann, J. (1979a). The influence of ammonia in nutrient solution on growth and metabolism of cucumber plants. *Pl Soil* **52**, 403–414.

Schenk, M. and Wehrmann, J. (1979b). Potassium and phosphate uptake of cucumber plants at different ammonia supply. *Pl Soil* **52**, 415–426.

Skinner, J. J. (1941). Plant-nutrient deficiencies in vegetable or truck-crop plants. In G. Hambidge, ed. Hunger Signs in Crops, 149–189. *Am Soc Agron Nat Fert Assoc Washington.*

Sonneveld, C. (1981). Irrigation and nutrition of glasshouse cucumbers in the Netherlands. *Proc 1981 Ann Western Greenhouse Veg Gro Conf Fresno Calif 1981*, 1–14.

Sonneveld, C. and Beusekom, J. van, (1974). The effect of saline irrigation water on some vegetables under glass. *Acta Hort* **35**, 75–85.

Sonneveld, C. and Voogt, S. J. (1975). Peat substrate as a growing medium for cucumbers. *Acta Hort* **50**, 45–52.

Sonneveld, C. and Voogt, S. J. (1978). Effects of saline irrigation water on glasshouse cucumbers. *Pl Soil* **49**, 595–606.

Von Stieglitz, C. R. and Chippendale, F. (1955). Nutritional disorders of plants. *Qd. Dep Agric Stk.* 36 pp.

Ward, G. M. (1967). Greenhouse cucumber nutrition: A growth analysis study. *Pl Soil* **26**, 324–332.

Ward, G. M. (1973a). Leaf analysis for greenhouse vegetable crops. *Ontario Minist Agric Fd* AGDEX Factsheet **290**.

Ward, G. M. (1973b). Calcium deficiency symptoms in greenhouse cucumbers. *Can J Pl Sci* **53**, 849–856.

Ward, G. M. (1976). Sulphur deficiency and toxicity symptoms in greenhouse tomatoes and cucumbers. *Can J Pl Sci* **56**, 133–137.

Wetzold, P. (1972). Ernährungsstörungen an Salatgurkenpflanzen Diagnose und Abhilfe. *Gemüse* **2**, 35–46.

CHAPTER 3
Lettuce

Lettuce *(Lactuca sativa L.)* is an important glasshouse crop, particularly during winter and early spring. Some specialist growers produce glasshouse lettuce, preferably under contract, on a year-round basis. The crop is more widely grown in the winter months, however, when outdoor conditions are unfavourable, in rotation with a high value summer crop such as tomatoes. The nutrient content of the soil is therefore determined to a large extent by the concentration of nutrients remaining from the preceding summer crop. Supplementary dressings of fertiliser can be applied before planting if necessary, but there is rarely time to reduce the concentration of soluble salts appreciably by leaching if excessive levels have accumulated. Liquid feeding is not practicably except for crops grown in specialized nutrient film systems e.g. Lauder, 1977. There is thus little opportunity for control of nutrition during cropping, and particular attention must therefore be given to this matter before planting. Recommendations for glasshouse lettuce in the U.K. have been published by MAFF (1982). Soil-grown lettuces are sensitive to acidity, possibly because of their susceptibility to manganese toxicity, and pH values of 6.5–7.0 are usually recommended.

Nitrogen

Symptoms and growth responses. Nitrogen deficiency decreased the growth and delayed the hearting of lettuce in sand culture. The leaves were pale green and eventually golden-yellow in colour (Woodman, 1939). Deficient plants remained small and pale green whilst the older leaves became yellow and withered (Messing, 1954). Maturity is delayed by N deficiency, and the plants often have exceptionally well developed roots (Roorda van Eysinga and Smilde, 1971). With a severe deficiency the plants are stunted, no heart is formed and the leaves may develop a purple or brown flush. Plants grown in sand culture with only NO_3-N were larger, firmer, flatter and less dark green in colour than those receiving NH_4-N only (Jager *et al.*, 1970). The adverse effect of NH_4-N was worst at low light intensity. When appreciable amounts of NH_4-N were added to soil, the first two leaves of the plants became yellow shortly after emergence and remained so for 28–32 days, during which time very little growth occurred (Paul and Polle, 1965).

The weight of lettuce produced increased with the concentration of N applied when the concentrations were low (Woodman, 1939), but was depressed at very high concentrations (Massey, 1975). In soil, N deficiency during head formation reduced the yield of autumn lettuce (Gardner and Pew, 1972), but applied N sometimes had little effect on the final yield (Haworth and Cleaver, 1967). Maximum yields are usually attained at intermediate levels of N (Winsor and Adams, 1968a; see Table 6). Heavy applications of N generally reduce the yield as, for example, in two years out of three with autumn lettuce (Gardner and Pew, 1972) and in one year out of three with spring lettuce (Gardner and Pew, 1974). The adverse effect of heavy application of N fertiliser was most marked in unlimed soil, being partly, though not entirely, due to acidity (Winsor, 1969). In soil with a low N content, the highest rate of applied N increased the yield by 10% whereas the same rate of fertiliser application depressed the yield by 3% on soil with a high N content (Roorda van Eysinga, 1966).

In a sphagnum peat soil, the yield of lettuce was restricted at low levels of applied N (MacKay and Chipman, 1961). Plants grown without added N in beds of peat were smaller and paler than those growing with an adequate supply; this was particularly noticeable in a crop planted under low-light conditions in December (Adams *et al.*, 1978a; see Fig. 6). Heavy applications of N sometimes depressed the yield (MacKay and Chipman, 1961), especially at low pH (4.0), but in well limed peat (pH 6.7) there was always a positive response to added N (Borkowski, 1978).

Nitrogen deficiency depressed the proportion of marketable lettuce by 50% and 29% in autumn and winter crops, but by only 3% in spring (Adams *et al.*, 1978a).

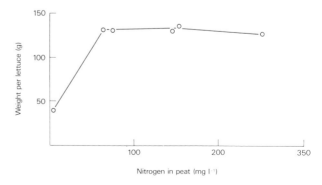

Figure 6 Relation between the inorganic nitrogen content of a peat substrate and the weight per lettuce (after Adams, *et al.*, 1978a).

Heavy dressings of N applied to unlimed soil also reduced the proportion of marketable lettuce, the response being most marked when the P status of the soil was inadequate (Winsor, 1969). A similar interaction between N and pH was found in peat (Adams *et al.*, 1978a). In both peat and soil the adverse effect of high levels of N is frequently associated with a depression of the pH and the release of excessive amounts of available Mn (see section on manganese pp 34, 35).

In sand culture, ammonium sulphate generally gave more sharply defined response curves and a lower optimal N concentration than was found with ammonium nitrate as the N source (Massey and Winsor, 1974; Massey, 1975). When comparing sources in solution culture, Ikeda and Osawa (1980) found far less inhibition of lettuce growth by NH_4-N than was the case for other crops such as spinach or Chinese cabbage. Growth was, however, markedly depressed by NO_2-N. A similar depression with NO_2-N was found in soil, but high levels of NH_4-N also caused a marked reduction in yield (Paul and Polle, 1964).

In trials at Fairfield Experimental Horticulture Station (Anon, 1964), ammonium sulphate and urea proved to be satisfactory sources of N at a low level of application (112 kg ha^{-1} N). A high rate of N application (448 kg ha^{-1}) depressed marketable yields by 24% and 19% respectively, but even larger depressions resulted where ammonium nitrate (28%) and calcium nitrate (57%) were applied at this level. The poor performance of the nitrate fertilisers in this trial may perhaps have been associated with soil salinity.

Nitrogen content. The total N content (%) was highest when the plants were young, and generally decreased with age (Slater and Goodall, 1957; Zink and Yamaguchi, 1962; Knavel, 1981). Nitrogen content was also highest in the young (heart) leaves and lowest in the outer leaves, the mean values reported by Knavel (1974) being 5.1% N and 3.5% N respectively. Knavel (1981) subsequently associated good growth of lettuce in sand culture with the relatively high levels of 5.4–5.7% N in the leaves of plants 60 days old. Analysis of young (25 day) lettuce plants may not be very helpful for diagnostic purposes, since little difference in content was found between normal and 'deficient' lettuce (Perez Melian *et al.*, 1977). In mature plants, however, these authors found that 3.0% N and 3.6% N were associated with severe and mild deficiencies respectively compared with 4.1% N in normal plants. Haworth and Cleaver (1967) recorded little response in their field trials over the range 35–104 kg ha^{-1} N; the plants contained 4.3–4.5% N at the last date of sampling, suggesting that even the lowest rate of application was fully adequate at this site. Cantliffe and Phatak (1974) reported that maximum weight of greenhouse lettuce grown in a soil:peat:sand mixture corresponded to 3.8% N in the tissue.

	K1	K2	K3	Mean
N1	44	74	44	54
N2	44	152	170	122
N3	50	128	160	113
Mean	46	118	126	

Table 6 Effect of nitrogen and potassium on the weight (g) per lettuce (cv. Delta; after Winsor and Adams, 1968). LSD (P = 0.05) between N and K means = 8.

Maximum growth of lettuce in outdoor plots of sphagnum peat with fertiliser applied at 0–270 kg ha^{-1} N occurred at 180 kg ha^{-1}, corresponding to 1.92% N in the midribs of the wrapper leaves or about 5.1% N in the laminae (Bishop *et al.*, 1973). With greenhouse lettuce in peat the yields increased until the leaves contained 4.9–5.0% N in autumn crops whereas 3.9% N sufficed for a later crop which was harvested in better light. (Adams *et al.*, 1978a). The corresponding values associated with N deficiency were 2.5% and 1.5% N respectively.

Slater and Goodall (1957) showed an interaction between the P status of lettuce grown in sand culture and their subsequent response to N. Thus the N content of the plants (44 days from sowing) above which no further positive response could be expected, increased progressively from 2.2% N at low P (0.1% P) to 3.6% N at high P (0.4% P).

Both the content of the leaves and the rate of uptake are influenced by temperature. Thus Knavel (1974) found that the N content of the leaves decreased from 5.0% to 4.4.% as the minimum air temperature increased from $-1°$ to 10°C. Frota and Tucker (1972) showed that the rates of uptake both of NH_4-N and of NO_3-N increased with temperature, though the response was greater with NO_3-N. Thus the uptake of NH_4-N predominated at low temperatures (2°–13°C); the rates were about equal at 18°C, and uptake of NO_3-N predominated at 23°C.

The NO_3-N content of the leaves decreased as the light intensity increased and therefore depended partly on the date of harvest (Roorda van Eysinga, 1966); it was, however, virtually unaffected by leaf age (Zink and Yamaguchi, 1962). A lower critical level of 0.25% NO_3-N has been proposed (Zink and Yamaguchi, 1962). Roorda van Eysinga and Smilde (1971) found 0.04% NO_3-N in deficient plants, whilst maximum yields were associated with 1.1–1.2% NO_3-N (Roorda van Eysinga, 1966) and 1.0–2.0% NO_3-N (Roorda van Eysinga and Smilde, 1971). Even when grown at a relatively low concentration (28 mg l^{-1} NO_3-N) in solution culture the leaves of lettuce plants contained over 1.0% NO_3-N (Ikeda and Osawa, 1980). Much attention has been given to this topic because of the possible health hazard associated with a high level of NO_3-N in human food; see, for example, Roorda van Eysinga (1984).

Phosphorus

Symptoms and growth responses. The leaves of P deficient plants are dull and dark green in colour. The plants fail to form a heart and the older leaves die prematurely (Messing, 1954). The plants are stunted but there are no characteristic symptoms of deficiency, although red or purple tints develop on the leaves of some cultivars (Roorda van Eysinga and Smilde, 1971; Perez Melian *et al.*, 1977).

The average weight per lettuce increased progressively with the rate of fertiliser P applied to unsteamed soil (Winsor and Adams, 1968b; Smith and Scaife, 1973). At low levels of P, growth was better in acid soil (pH 5.1), but liming to pH 6.9 was beneficial where heavy dressings of P had been applied (Winsor and Adams, 1968b). Where steam sterilisation releases appreciable amounts of Mn, however, liming can be beneficial even at low P levels. The

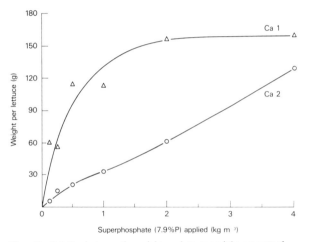

Figure 7 Relation between the weight per lettuce and the amount of superphosphate applied for the preceding cucumber crop at two levels of liming; pH 6.0 and 7.2 at Ca 1 and Ca 2 respectively (after Adams and Winsor, 1984).

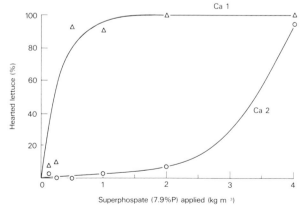

Figure 8 Relation between the proportion of hearted lettuce (%) and the amount of superphosphate applied for the preceding cucumber crop at two levels of liming; pH 6.0 and 7.2 at Ca 1 and Ca 2 respectively (after Adams and Winsor, 1984).

pH	Superphosphate applied to preceding crop (kg m^{-3})						Mean
	0.125	0.25	0.5	1.0	2.0	4.0	
6.0	43	60	100	100	100	100	84
7.2	2	10	12	31	71	100	38
Mean	23	35	56	66	86	100	

Table 7 Effects of phosphorus and pH on the proportion (%) of marketable lettuce grown in peat (unpublished data, Adams and Winsor).

yield response to applied P increased progressively with the level of K (Winsor and Long, 1963a). High P helped to mitigate the adverse effects of excess N on yield (Winsor and Long, 1961; Mondrzak-Rosenberg, 1965). Application of P increased early yields more than the final yields, indicating that high levels promote earlier maturity (Lorenz *et al.*, 1964). Thus the earliest maturity occurred on plots receiving 235 kg ha^{-1} P or more, whereas there was no further increase in total yield above 58 kg ha^{-1} P. On newly cultivated organic soils, heavy dressings of P were required to achieve maximum yield (Parups and Goodwin-Wilson, 1958). In peat, the yield increased with the rate of superphosphate (7.9% P) applied (to the preceding cucumber crop) up to 4 kg m^{-3} at pH 7.2, but there was no response to dressings of more than 2 kg m^{-3} at pH 6.0 (Adams and Winsor, 1984; see Fig. 7). The response to P was very dependent on pH; the higher pH (7.2) depressed the yield severely at low and intermediate levels of applied phosphate.

Addition of phosphate to a P deficient glasshouse soil increased the proportion of marketable (hearted) lettuce from 24% to 85% overall (P <0.001) and from 5% to 81% in the unlimed plots (Winsor *et al.*, 1964, with further unpublished data). In peat, the response to liming was also influenced by pH, though the form of the interaction differed from that observed in the soil. For example, at pH 6.0 all the lettuce were marketable where 0.5 kg m^{-3} superphosphate had been applied for the preceding crop whereas at pH 7.2 only 12% were marketable, and the residual P from heavy dressings of superphosphate (4 kg m^{-3}) was necessary to ensure that all plants were marketable (Adams and Winsor, unpublished data; see Table 7). The effects of P and pH on the hearting of lettuce grown in peat are shown in Figure 8.

Phosphorus content of the plants. The P content of the leaves increased with the level of P applied, and generally declined with increasing age of the plants (Grant Lipp and Goodall, 1958a). No positive response to added P would be expected at tissue levels above 0.47% P at 29 days or 0.37% P at 37–44 days. Pandita and Andrew (1967) found that the P content of the lettuce declined markedly during the first eight weeks after germination, but only slightly thereafter. Liming depressed the P content of plants adequately supplied with P but not of deficient plants (Winsor and Long, 1963b). Potassium deficient lettuces had a higher P content than normal ones (Table 8) and, where an adequate amount of P was applied, heavy dressings of K resulted in a small decrease in P content (Winsor and Long, 1963b).

In peat, water soluble P declined markedly between pH 5.6 and 6.5, but the P content of the leaves remained practically constant until the pH exceeded 6.5 (Adams *et al.*, 1978a; see Fig. 9).

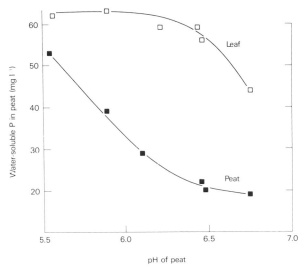

Figure 9 Effect of pH on the water-soluble phosphorus content of the peat and on the phosphorus status of lettuce (after Adams *et al.*, 1978a).

The dry weight of young lettuce was closely correlated (r = 0.91) with the P content of the leaves. In older plants, however, the contents fell within a more restricted range and were less well correlated with plant weight (Smith and Scaife, 1973). Lettuce with a higher P content matured earlier in spring (Pandita and Andrew, 1967), though Roorda van Eysinga and Smilde (1971) reported that Dutch cultivars showed no such relation. The youngest leaves had a higher P content (0.67% P) than the older leaves (0.41% P), but minimum temperatures of between $-1°$ and $10°C$ had little effect on the levels (Knavel, 1974).

Published values for the P content of healthy lettuce include 0.34–0.74% with the majority of the samples in the range 0.4–0.6% (Zink and Yamaguchi, 1962), 0.78–1.31% (Roorda van Eysinga and Smilde, 1971), 0.6% (Knavel, 1974), 0.4–0.5% for mature and young plants respectively (Perez Melian *et al.*, 1977) and 0.6–0.8% (Adams and Winsor, 1984). Berry *et al.* (1981) found 0.55–0.76% P (average 0.65% P) in lettuce grown for 28 days in a peat/vermiculite substrate in controlled environment chambers at five different laboratories.

Severe deficiency in young plants has been associated with values of about 0.09% P (Grant Lipp and Goodall,

| | Potassium level | | | |
	K1	K2	K3	Mean
Unlimed	0.96	0.83	0.82	0.87
Limed	0.92	0.72	0.71	0.78
Mean	0.94	0.78	0.77	

Table 8 Effects of potassium level and liming on the phosphorus content (% P) of winter lettuce grown in a soil adequately supplied with fertiliser phosphorus (after Winsor and Long, 1963b).

1958a) and 0.2% P (Perez Melian *et al.*, 1977). Berry (1971) related the growth of lettuce seedlings in solution culture for 13 days (17 days from germination) to the P content of the tissue (% P soluble in 2% acetic acid, expressed on the basis of dry matter). Critical levels, corresponding to 10% growth reduction, were interpolated as 0.078% P in the laminae and 0.060% P in the conductive tissue (mid-rib and petiole); a slightly higher 'diagnostic level' of 0.1% P in the tops of the lettuce seedlings was regarded as preferable. A critical level of 0.2% P (soluble in 2% acetic acid) in the mid-rib tissue was given by Tyler and Lorenz (1962) and Lorenz *et al.* (1964).

Sulphur

Sulphur deficiency is apparently unknown in commercial glasshouses owing to the use of SO_4 containing fertilisers such as superphosphate, potassium sulphate and magnesium sulphate. The condition was illustrated by Roorda van Eysinga and Smilde (1971), based on experiments in water culture. Affected plants had an overall yellow-green colour and growth was restricted. As noted for certain other crops, the leaves of S deficient plants are stiffer than normal. A very similar description of a young plant grown in sand culture was given by Scaife and Turner (1983). Mean analytical values of 0.29% total-S and 0.13% SO_4-S in healthy lettuce plants were reported by Roorda van Eysinga and Smilde (1981).

Potassium

Symptoms and growth responses. An inadequate supply of K restricts growth and the leaves become a darker green than those of normal plants. With a severe deficiency, yellow spots develop near the tips of the older leaves. These spots spread, coalesce, and may eventually become brown (Roorda van Eysinga and Smilde, 1971). In sand culture, there was no noticeable response to the omission of K for three weeks after the treatment was applied. Then the growth of the new leaves ceased and the older leaves developed a marginal scorch, turned yellow and died (Messing, 1954): see also the illustration by Wallace (1951,

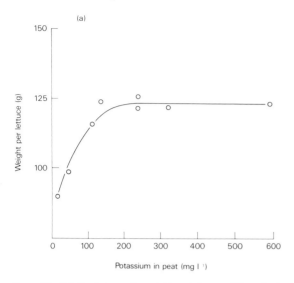

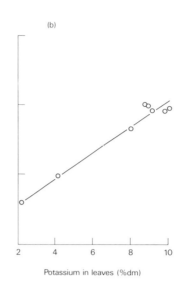

Figure 10 Relation between the weight per lettuce and the potassium content of (a) the peat substrate and (b) the leaves (after Adams *et al.*, 1978a).

1961). Goodall *et al.* (1955) noted a dull, dark green colour in deficient lettuce, and also recorded that the leaves were thick, rounded, smoother than normal and with flat margins; the root systems were poorly developed and the plants failed to heart. Plants grown in peat at very low K levels were reduced in size, but specific leaf symptoms did not develop until the crop approached maturity. The outer leaves then became yellow and, particularly in spells of bright weather, wilted and died surprisingly quickly (Adams *et al.*, 1978a).

Winsor and Long (1963a) found a marked response of winter lettuce to residual levels of K applied to preceding tomato crops (for fertiliser treatments see Winsor *et al.*, 1967); the weight per lettuce was increased by 58% and 89% respectively at the intermediate and high levels where P was not limiting. As shown in Table 6, very high residual concentrations of K in the soil were favourable to growth except when N deficiency was limiting (Winsor and Adams, 1968a; Adams and Winsor, 1973).

In peat, the yield of field-grown lettuce increased only up to intermediate rates of application (150 kg ha^{-1} K; Bishop *et al.*, 1973). Borkowski *et al.* (1975) found optimum growth of greenhouse lettuce in peat with a basal dressing of 500–800 mg l^{-1} K; slight depression in yield was recorded above 1000–1200 mg l^{-1} K. The growth response of glasshouse lettuce to increasing levels of K (Fig. 10a) was very marked at low concentrations of extractable K in the peat, but a moderate concentration sufficed for maximum yield (Adams *et al.*, 1978a).

In a factorial study of the nutrition of lettuce in soil the proportion of marketable (hearted) lettuce in plots receiving adequate N and P increased from 22% at low K to 97% and 100% respectively at intermediate and high K levels

(Winsor and Adams, 1968b and unpublished data). In peat, plants grown at very low concentrations of extractable K developed a marginal leaf scorch which rendered many of them unmarketable. The proportion of marketable lettuce increased to a maximum at quite low concentrations of K (100–150 mg l^{-1} K extractable with ammonium acetate-acetic acid) with little change at higher concentrations (<600 mg l^{-1} K; Adams *et al.*, 1978a).

Potassium content of the plants. Various authors have recorded the increased K content of lettuce plants in relation to K application. Thus Lambeth (1953) found 2.7–10.6% K in plants grown at the more favourable of two levels of N applied (112 kg ha^{-1} N); Grant Lipp and Goodall (1958b) found 1.2–4.0% K in young lettuce in sand culture at five levels of K 44 days after sowing, and Winsor (1968) recorded 3.4–8.8% K for mature lettuce grown in soil at three levels of K. The dry weight of lettuce seedlings increased with the content of the leaves up to 4% K, but there was no clear relationship between dry weight and content between 4% and 8% K (Berry and Carey, 1971); the authors proposed a critical level of 2% K, corresponding to a 10% reduction in yield. In peat, the yield increased with the K content of the leaves up to 8–9% see Figure 10b. The response was more marked in a crop planted in September than in those planted in October or December (Adams *et al.*, 1978a).

The K content of the youngest leaves (4.39–4.45% K) was unaffected by changes in minimum temperature between −1° and 10°C, but the lower temperature resulted in a higher content (5.6% K) in the older leaves (Knavel, 1974).

A wide range of K contents have been associated with good growth e.g. 4.6–8.4% K (Zink and Yamaguchi, 1962), 4.1–10.0% (Roorda van Eysinga and Smilde, 1971), 7.9–9.4% (Bishop *et al.*, 1973), 8.0–9.0% (Adams *et al.*, 1978a) and 9–10% (Knavel, 1981). Potassium deficiency is associated with contents of less than 2% K, corresponding to 10% yield reduction in young plants (Berry and Carey, 1971), 0.75–2.07%, with growth reduction below 4.2% K (Roorda van Eysinga and Smilde, 1971) and 2.2% K (Adams *et al.*, 1978a). Perez Melian *et al.* (1977) reported 6.3–7.2% K corresponding to about 15% growth reduction and 1.8–2.2% K corresponding to 30% growth reduction, without deficiency symptoms. The same authors found 9.1% K in the leaves when excessive levels of K were supplied (430 mg l^{-1} K in solution, used in aggregate culture), with a maximum of 9.4% K where Ca or N were deficient.

Calcium

Symptoms and growth responses. The margins of the young leaves of Ca deficient plants were distorted and scorched (Wallace, 1951). Dark brown, almost black, spots developed on the youngest leaves; the growing point died and the spots spread to the older leaves, which died off in succession (Messing, 1954). The irregularly shaped dark lesions first appeared at the periphery of the basal part of the leaf, later spreading over the entire leaf surface and frequently coalescing (Struckmeyer and Tibbitts, 1965). The epidermal and mesophyll cells in the affected areas collapsed, as also did cells within the vascular bundles. The xylem elements became occluded with a gum-like substance. Growth was severely restricted, particularly in the young leaves which were somewhat darker and more crinkled than normal (Roorda van Eysinga and Smilde, 1971).

Much has been written on the possible role of Ca in the disorder known as 'tipburn'. As with other disorders relating to this deficiency e.g. blossom-end rot of tomatoes, the relationship is rarely a simple one. Struckmeyer and Tibbitts (1965) concluded that the internal symptoms of Ca deficiency were not comparable with those described for tipburn (Tibbits *et al.*, 1965), since no extrusion of latex occurred from deficient plants and the laticifers remained normal. Several different forms of tipburn may be distinguished. Thus Borkowski (1978) showed opposing responses of 'dry tipburn' and 'latex tipburn' to progressive applications of calcium carbonate on an acid peat; the incidence of dry tipburn was decreased by liming from pH 5.0 to pH 6.2, whilst that of latex tipburn increased significantly over the same range. Foliar sprays of calcium nitrate or chloride prevented tipburn, provided that the treatment reached the heart leaves (Thibodeau and Minotti, 1969). It was suggested that lettuce tipburn was associated with a temporary and localised shortage of soluble Ca during a period of high demand, as during rapid growth; see also later studies by Collier and Huntingdon

(1983). Practical control of tipburn in field-grown lettuce by Ca sprays is unlikely, however, owing to the inaccessibility of the immature leaves within the heads. Ashkar and Ries (1971) showed that tipburn was readily induced by a low Ca and high NO_3-N regime; the disorder was associated with an accumulation of free amino acids, particularly aspartic and glutamic acids and their amides. Collier and Holness (1984) showed that spraying lettuce plants with either ammonium oxalate or with chelating agents increased the incidence of tipburn. It was concluded that lowering the concentration of Ca^{++} but not the total Ca concentration can induce the disorder. Calcium concentrations were increased, and the incidence of tipburn delayed, by humidity conditions which provided large diurnal fluctuations in water potential in the plant and which encouraged root pressure flow during the dark period (Collier and Tibbitts, 1984).

In sand culture, the plant weight increased with the concentration of Ca (up to 20 mM) supplied in the nutrient solution, this response being most marked at a high level of applied NO_3-N (Ashkar and Ries, 1971). Slight Ca deficiency reduced the weight per lettuce by 14% whereas a severe deficiency depressed plant weight by 30% and resulted in death of the younger leaves, thus rendering the lettuce unsaleable (Perez Melian *et al.*, 1977).

In glasshouse soil, liming increased the weight per lettuce, particularly where heavy dressings of N had been applied (Winsor and Long, 1963a). Again, growth was improved by liming, even when the average pH increased from 6.2 to 7.7 (Winsor and Adams, 1968b). Much of this response may have resulted from a reduction in the available Mn content of the soil which, due to the soil type (brick-earth), is released in appreciable amounts after partial steam-sterilisation. Nevertheless, when the virgin soil was adequately supplied with P but not steamed, the lettuce grew better after a heavy application of lime (pH 6.9) than in the unlimed soil (pH 5.1; Winsor and Adams, 1968a). A very marked increase in plant weight was also found as peat was limed progressively from pH 3.7 to 6.2 (Borkowski, 1977), although a depression in yield due to liming peat of low P status has also been observed (Adams *et al.*, 1975).

Calcium content of the plants. Increasing the Ca concentration in the nutrient solution over the range 0–800 mg l^{-1} Ca increased the content of the leaves from 0.8% to 2.9% Ca. The concomitant depression in the P content of the leaves with increasing Ca levels (1.2% to 0.5% P at a low N level and from 0.7% to 0.2% P at high N) did not prevent increases of 20% and 41% respectively in plant weight (Ashkar and Ries, 1971). Increasing the minimum temperature from $-1°$ to 10°C increased the calcium content of the oldest leaves from 1.35% Ca to 1.72% Ca (P <0.05; Knavel, 1974). Far lower calcium contents (0.26–0.33% Ca) were found in the youngest leaves. Applications of N and K depressed the Ca content

(Davidson and Thiegs, 1966), and unbalanced high N increased the susceptibility of the lettuce to *Botrytis cinerea* (Krauss, 1971). Increasing the concentration of Ca in the nutrient solution increased the Ca content far more markedly at high light than at low light intensities (Ashkar and Ries, 1971).

Published values of Ca contents associated with good growth of lettuce vary somewhat, including 1.0–1.5% Ca (Zink and Yamaguchi, 1962), 1.0–1.3% Ca (Haworth and Cleaver, 1967), 1.1–1.8% Ca (Roorda van Eysinga and Smilde, 1971), 1.6–3.0% Ca (Ashkar and Ries, 1971), 0.9–1.1% Ca (Perez Melian *et al.*, 1977), 1.2–1.6% Ca (Knavel, 1981) and 1.2–1.7% Ca (Berry *et al.*, 1981). Amongst these values, the extreme figure of 3.0% Ca reported by Ashkar and Ries (1971) was for plants grown in vermiculite at a very high concentration (800 mg l^{-1} Ca); the fresh weight of the plants at this level exceeded that of plants grown at 200 mg l^{-1} Ca. Lorenz and Minges (1942) found a progressive decline in the Ca content of lettuce plants during growth in the field, the values ranging from 2.55% Ca in July to 1.19% Ca in September.

Severely deficient plants contain 0.2–0.6% Ca (Roorda van Eysinga and Smilde, 1971) and 0.4% Ca (Perez Melian *et al.*, 1977) whilst those with no symptoms except retarded growth contained 0.9–1.3% Ca (Roorda van Eysinga and Smilde, 1971) and 0.9% Ca (Perez Melian *et al.*, 1977).

Magnesium

Symptoms and growth responses. The growth rate was greatly reduced and hearting was prevented by Mg deficiency. Interveinal and marginal yellowing developed on the leaves, and the margins of the oldest leaves eventually became scorched (Messing, 1954; see also Wallace, 1951). The yellowing may begin on the oldest leaves and spread interveinally from the margins although, in water culture, the symptoms develop first on the younger leaves (Roorda van Eysinga and Smilde, 1971). In soil where Mg had been omitted from the base fertiliser dressings for ten years, lettuce did not develop symptoms of deficiency, despite the marked deficiency symptoms and significant loss in yield found in both the preceding and subsequent tomato crops (Winsor and Adams, 1968b). A similar lack of symptoms was noted with lettuce grown between responsive tomato crops in peat (Adams *et al.*, 1978a). Magnesium treatment had no significant effect on the yield and proportion of marketable lettuce in these experiments either in soil or in peat. A similar lack of response to Mg by lettuce grown in soil was noted by Lambeth (1953), though Marlatt (1959) reported a significant increase in the head weights of field-grown lettuce receiving chelated magnesium (EDTA) at rates of 0.6–3.1 kg ha^{-1} Mg.

Magnesium content of the plants. The Mg content of the leaves was depressed by increasing levels of K (e.g. Lambeth, 1953; Borkowski *et al.*, 1975) and by increasing the concentration of Ca in the nutrient solution; the latter response was particularly marked at a high level of applied N and depressed the Mg content from 1.2% to 0.6% Mg (Ashkar and Ries, 1971). The Mg content of the leaves also declines somewhat as the plants approach maturity (Zink and Yamaguchi, 1962).

The Mg content of lettuce plants is depressed by increasing levels of K. Thus Lambeth (1953), growing lettuce in soils fertilised to different levels of K saturation (2.7–8.1%), reported 0.84% Mg in plants at low K compared with 0.40–0.53% Mg at intermediate and high levels. Similarly, Borkowski *et al.* (1975) found 0.77–0.92% Mg at low K compared with 0.53–0.75% Mg at high K. Haworth and Cleaver (1967) found a lower Mg content (0.41% Mg) in lettuce from plots receiving dung than from untreated plots (0.62% Mg); this response may well be associated with the higher K content of the plants receiving dung. The Mg content of lettuce grown in vermiculite (Ashkar and Ries, 1971) also decreased with increasing Ca concentration, the mean values being 1.2%, 0.8% and 0.75% respectively at solution concentrations of 0, 5 and 20 mM Ca. Magnesium contents reported by Zink and Yamaguchi (1962) from 17 field trials in California were in the range 0.40–0.56% Mg with a mean of 0.47% Mg; the extreme values for individual samples were 0.32% and 0.84% Mg.

Roorda van Eysinga and Smilde (1971) recorded 0.24–0.48% Mg in the heads of healthy lettuce; severe deficiency symptoms and stunted growth were associated with 0.05–0.10% Mg. The same authors (1981) quoted a range of 0.37–0.90% Mg for healthy lettuce, with deficiency below 0.29% Mg. The Mg content was higher in the outer leaves (0.52% Mg) than in the inner leaves (0.30% Mg; Knavel, 1974). Perez Melian *et al.* (1977) found higher Mg levels in K deficient lettuce, 1.0% and 0.7% Mg in young and mature lettuce respectively, than in plants treated with excess Mg (corresponding values of 0.5% and 0.4% Mg) or in plants receiving the standard nutrient solution (0.4% and 0.2% for young and mature respectively). Berry *et al.* (1981) recorded 0.27–0.39% Mg in lettuce grown under standardised conditions in controlled environmental chambers at five sites. Data reported by Knavel (1981) show an increase in the Mg content of lettuce with increasing night temperatures; thus mean values of 0.60%, 0.70% and 0.81% Mg were found for temperatures of 7°, 13° and 18° respectively.

Iron

Symptoms and growth responses. Deficient plants become pale green and the growth is retarded. The young leaves become yellow, the oldest leaves die and growth ceases (Roorda van Eysinga and Smilde, 1971). The veins of the older leaves were less chlorotic than the interveinal tissue, but when at a later stage the youngest leaves turned pale yellow the veins became chlorotic too. In his key to non-pathogenic disorders of lettuce, however, Marlatt (1967) stressed that the veins of Fe deficient plants remained distinctly green.

Despite the importance of Fe as an essential nutrient, few responses of lettuce have been recorded to variations in the level of applied Fe. In sand culture, Harward et al. (1955) found a progressive depression in the yield of lettuce supplied with increasing levels of ferric citrate over the range 0–25 mg l^{-1} Fe.

Moore et al. (1957) concluded that, with both NH_4-N (25%) and NO_3-N (75%) present in the nutrient solution and Fe supplied in the Fe^{++} form, maximum yield of lettuce corresponded to the relatively low concentration of 0.079 mg l^{-1} Fe (with Cu at 0.0026 mg l^{-1}). A higher concentration of 0.28 mg l^{-1} Fe (with 0.0065 mg l^{-1} Cu) was required when supplied in the ferric form. Marked interactions were found between Cu and Fe.

Iron content of the plants. Twyman (1951) recorded higher levels of Fe in Mn deficient than in normal lettuce. For example, when grown with 2 mg l^{-1} Fe in the solutions, plants showing symptoms of Mn deficiency contained 331 µg g^{-1} Fe compared with 236 µg g^{-1} Fe in plants adequately supplied with Mn. The concentration of Fe in the tissue bore no relation to the incidence of chlorosis and, where supply limited growth and development, this element sometimes accumulated to quite high levels. The Fe content of lettuce plants grown in sand culture increased with the concentration of Fe supplied, from 37 µg g^{-1} Fe at 0 and 0.2 mg l^{-1} Fe to 80 µg g^{-1} at 25 mg l^{-1} Fe, accompanied by a progressive depression of Cu content in the leaves (Harward et al., 1955). The Fe content was higher in the old than in the young leaves, particularly at high levels of supply. Moore et al. (1957) reported Fe contents ranging from 30 µg g^{-1} to 632 µg g^{-1} for lettuce grown in solution culture. The average content of the plants was greater with Fe^{++} than with Fe^{+++}, and was also higher when part (25%) of the N was supplied as NH_4-N compared with solutions containing only NO_3-N. Lettuce plants with distinct symptoms of Fe deficiency contained 93–604 µg g^{-1} Fe whereas 130–1468 µg g^{-1} Fe were recorded for healthy plants (Roorda van Eysinga and Smilde, 1971); the range for deficient and normal plants thus overlapped considerably, and both included some surprisingly high values. Berry et al. (1981), in a collaborative study at five centres using a standard peat-vermiculite substrate, found 126–276 µg g^{-1} Fe in the leaves, with a mean value of 202 µg g^{-1}.

It is clear that the total Fe content of plant tissue is of limited value for diagnostic purposes. Low values, for example below about 50 µg g^{-1} Fe, suggest inadequate supply or uptake, though higher values in no way exclude

the possibility of Fe deficiency owing to inactivation in the tissue, as already noted by Twyman (1951).

Manganese

Symptoms of deficiency. Chlorosis of Mn deficient lettuce was recorded by Gilbert and McLean (1928), accompanied by yield reductions in four out of five experimental field plots. The condition, described as 'chlorotic marbling', was illustrated by Wallace (1951, 1961); see also the illustration in volume 2 of the present publication (Scaife and Turner, 1983). Roorda van Eysinga and Smilde (1971) recorded an overall yellowish-green colour though growth was not greatly affected. With acute deficiency the older leaves in particular became chlorotic, but the veins, including even the smallest ones, remained green. Marlatt (1967) noted the chlorosis and also described the plants as stunted and having a grayish bloom on the underside of the leaves. The leaves were sometimes distorted, with hollow mid-ribs; irregular necrotic spots occurred along the ribs and small, well-defined, spots on the leaf margins.

Manganese content of deficient and normal lettuce. Cos lettuce with pronounced symptoms of Mn deficiency contained only 8 µg g^{-1} Mn in the dried tissue (Nicholas, 1949). Cabbage lettuce showing very slight deficiency contained 18 µg g^{-1}. Twyman (1951) noted that deficiency symptoms were always present when the aerial part of the plants contained less than 12 µg g^{-1} Mn. Manganese deficiency was associated with values of 6–14 µg g^{-1} Mn (Roorda van Eysinga and Smilde, 1971), 14 µg g^{-1} (Vlamis and Williams, 1973) and below 22 µg g^{-1} Mn (Roorda van Eysinga and Smilde, 1981).

Maximum growth of lettuce in sand culture (Harward et al., 1955) corresponded with 27 µg g^{-1} Mn in the young leaves and 104 µg g^{-1} in the older (outer) leaves; a higher content of Mn in the older tissue was also noted by Sonneveld and Voogt (1971a) and others. Massey and Winsor (1960) found 126–175 µg g^{-1} Mn in healthy lettuce grown in steam-sterilised glasshouse soil, and Roorda van Eysinga and Smilde (1981) recorded 30–198 µg g^{-1} Mn. Messing (1965b) observed the best growth of four lettuce cultivars in sand culture, with freedom from either deficiency or toxicity symptoms, at 1 mg l^{-1} Mn in solution, corresponding to 90–119 µg g^{-1} Mn in the plant tissue. Vlamis and Williams (1973) found maximum growth at 0.1–0.5 mg l^{-1} Mn in aerated solution culture. These treatments corresponded to 63–130 µg g^{-1} Mn in the young leaves, 89–204 µg g^{-1} in the older leaves after removal of the leaf edges and 307–704 µg g^{-1} in the leaf edges themselves. Berry et al. (1981) reported 14–136 µg g^{-1} Mn in young (28-day) lettuce plants grown under controlled conditions at five sites. The value of 14 µg g^{-1} at one of these collaborating research centres seems surprisingly low for plants grown with 0.25 mg l^{-1} Mn in solution; the range for the other four sites was 45–136 µg g^{-1} Mn.

Manganese toxicity. Manganese toxicity occurs on many glasshouse soils and loam-based potting mixtures after steam sterilisation, which can release large quantities of Mn in water-soluble and exchangeable form; see, for example, Davies (1957), Massey and Winsor (1960), Messing (1965a), Sonneveld (1968), Jager *et al.* (1969a, 1969b, 1970), Sonneveld and Voogt (1975a,b). Lettuce plants are particularly susceptible to damage by excess Mn, and this condition can occur even without sterilisation on acid soils.

Plant (1956) described and illustrated toxicity as characterized by a golden margin extending round all the leaves, though most noticeable in the lowest and largest ones. Williams and Vlamis (1957) described and illustrated Mn toxicity in lettuce (cv. Romaine) when grown in solution culture at 0.5 µg l^{-1} Mn; the symptoms consisted of small necrotic spots scattered over the surface of the older leaves. The same authors (Vlamis and Williams, 1973) subsequently described Mn toxicity in lettuce (cv. Romaine) in terms of marginal yellowing of the leaves; the chlorosis was absent at 0.5 mg l^{-1} Mn but distinct at 2 mg l^{-1} and severe at 5 and 10 mg l^{-1} Mn. Messing (1965a) grew lettuce in a steamed loam:peat (5:1) mixture with different amounts of lime and P added. The symptoms of Mn excess consisted of marginal chlorosis and scattered necrotic spots along the margins of the older leaves. These symptoms occurred in treatments having pH values of 4.8–5.5, accompanied by severe stunting, with milder symptoms at pH 5.6–6.0. Jager *et al.* (1970) induced symptoms of Mn excess in lettuce grown in sand culture. The oldest leaves developed fine brown necrotic spots over the whole surface, followed by yellowing and death of the leaf margins. A detailed description of Mn toxicity was given by Roorda van Eysinga and Smilde (1971); the leaf margins become yellow and dark brown spots develop along these margins, particularly near the tip. The spots coalesce, resulting in a marginal scorch. The older leaves die prematurely, the younger leaves are stunted, and the 'tulip-shaped' lettuce fails to form a heart.

Harward *et al.* (1955) observed symptoms of Mn toxicity in lettuce plants containing an average of 179 µg g^{-1} Mn in the young leaves and 549 µg g^{-1} in the older leaves. Massey and Winsor (1960) found that young plants containing 175 µg g^{-1} Mn showed no symptoms of the toxicity, those containing 350 µg g^{-1} had moderately well developed symptoms whilst those containing 600 µg g^{-1} Mn or more showed severe toxicity symptoms. Messing (1965a) reported mild toxicity symptoms at 200–500 µg g^{-1} Mn in the plants, with more severe symptoms above 500 µg g^{-1}. Roorda van Eysinga and Smilde (1971) regarded 200 µg g^{-1} Mn as an excessive content for lettuce, though distinct symptoms did not usually occur

until the leaves contained 300 µg g^{-1} or more. Vlamis and Williams (1973) found no foliar symptoms when the young leaves contained 130 µg g^{-1} Mn, but distinct symptoms occurred when these leaves contained 333 µg g^{-1} Mn.

Lettuce cultivars vary considerably in their susceptibility to Mn toxicity. Thus Messing (1965b), comparing four cultivars at a range of concentrations in sand culture, observed yield reductions of only 2% and 15% for two relatively tolerant cultivars grown with high Mn (16 mg l^{-1}) compared with 53–56% for two susceptible cultivars. Sonneveld and Voogt (1975b) recorded severe toxicity symptoms on one cultivar containing 432 µg g^{-1} Mn whereas the corresponding level for another cultivar was 979 µg g^{-1}.

The Mn contents of lettuce tissue associated with reduction of yields far exceed those required to induce visual symptoms of toxicity. Thus, whereas symptoms of toxicity were apparent at concentrations of 180 µg g^{-1} and 550 µg g^{-1} Mn in the young and old leaves respectively, no significant loss of yield resulted until levels of 350 µg g^{-1} and 930 µg g^{-1} Mn were reached (Harward *et al.*, 1955). Similarly, Messing (1965a) noted that, whereas symptoms of toxicity occurred at 200–500 µg g^{-1} Mn, the total yield in sand culture did not decrease until a tissue concentration of 1127 µg g^{-1} Mn was exceeded (cv. Cheshunt 5B). One cultivar ('Northern Queen') showed no loss in yield even at a concentration of 1459 µg g^{-1} Mn in the tissue (Messing, 1965b).

When lettuces (cv. Cheshunt 5B) were grown in an acid, steam-sterilised soil (Table 9), loss of yield was noted at a lower concentration of Mn in the tissue (47% yield reduction at 615 µg g^{-1} Mn) than was the case in sand culture (15% yield reduction at 1577 µg g^{-1} Mn). This may indicate the presence of some additional toxic factor in the acid soil, possibly aluminium (Messing, 1965a, 1971; see also Schmehl *et al.*, 1950). The relationship between

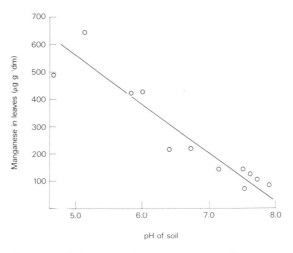

Figure 11   Relation between the manganese content of lettuce and the pH of the soil (after Messing, unpublished data).

Soil		Sand culture	
Mn content (μg g^{-1})	Relative yield (%)	Mn content (μg g^{-1})	Relative yield (%)
226	100	99	100
615	53	305	103
758	16	519	107
1306	9	1127	107
1867	1	1577	85

Table 9 Relation between yield and the manganese content of lettuce grown in soil and sand culture (cv. Cheshunt 5B; after Messing, 1965a).

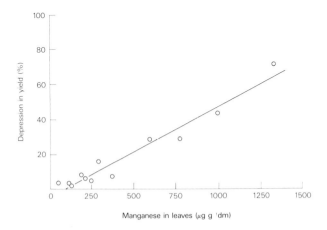

Figure 12 Relation between the manganese content of lettuce and the depression in yield (after Adams *et al.*, 1978a).

soil pH and the Mn content of lettuce is shown in Figure 11.

In peat containing a fritted micronutrient mixture, increasing levels of ammonium nitrate and of urea progressively depressed the pH and resulted in increasingly severe Mn toxicity (Adams *et al.*, 1978a). The loss in yield, expressed as a percentage of the yield from the most favourable treatment, was highly correlated (r = +0.97) with the content of the leaves in the range 200–1400 μg g^{-1} Mn (Fig. 12). The greatest loss in yield (75%) was associated with a pH of 5.2 whilst maximum yield was obtained at pH 6.4. There was also a negative linear relationship between the proportion of saleable lettuce and the Mn content of the leaves; less than 15% of the plants containing 1000 μg g^{-1} Mn were marketable.

Copper

Deficiency symptoms and growth responses. Copper deficiency is usually associated with highly organic soils and peat or peat-based substrates. Marlatt (1967) referred to

chlorosis of the leaf margins and petioles, sometimes extending over the whole leaf at a later stage. The leaves were stunted, narrow and limp. The condition was described and illustrated by Roorda van Eysinga and Smilde (1971); see also Smilde (1972). The leaves of deficient plants are narrow and cupped with some yellowing along the margins. The yellowing spreads across the leaves, the margins wilt, the veins become pink and the symptoms progress from the older to the younger leaves. Growth is reduced and the plants fail to form hearts. The leaves become cupped and twisted, giving the plants an untidy, whorled appearance (Adams *et al.*, 1978b).

MacKay *et al.* (1966) found yield increases of up to 18%, averaged over five successive crops, where Cu was applied to an acid peat soil. Omission of Cu from a sphagnum peat substrate limed to pH 5.1–6.1 reduced the dry matter production of young lettuce plants by over 30% (Smilde, 1972). A severe deficiency in peat (pH 6.8) reduced the fresh weight of the plants by 95% (Adams *et al.*, 1975), whilst a moderate deficiency depressed the yield by 53% and rendered 73% of the produce unmarketable (Adams *et al.*, 1978b).

Copper content of normal and deficient plants. Moore *et al.* (1957) concluded that the Cu content of the plants corresponding to maximum yield varied with the form (Fe^{++} or Fe^{+++}) and concentration of Fe supplied, the form of N (NO_3 or $NH_4 + NO_3$-N) and possibly also with Mo level. Examination of the response curves suggested that a concentration of about 5 μg g^{-1} Cu in the plants was generally favourable. MacKay *et al.* (1966) found 3.1–13.2 μg g^{-1} Cu in plants grown at six levels of Cu ranging from deficiency to excess; maximum yield (mean of five crops) corresponded to 7.8 μg g^{-1} Cu in the tissue.

Deficient plants contained less than 2 μg g^{-1} Cu as compared with 7–17 μg g^{-1} in healthy plants (Roorda van Eysinga and Smilde, 1971). Smilde (1972), working with peat-grown lettuce, concluded that growth responses to Cu application were normally associated with leaf concentrations below 3.6 μg g^{-1} Cu. Adams *et al.* (1978b) found 1.4–1.8 μg g^{-1} Cu in lettuce showing symptoms of deficiency compared with 7.4–8.7 μg g^{-1} in plants adequately supplied with this nutrient. No reduction in yield (fresh weight) was found at concentrations down to 2.4 μg g^{-1} Cu in the leaves, though the proportion of marketable lettuce was somewhat decreased at this level. Berry *et al.* (1981) recorded 6.0–13.8 μg g^{-1} Cu in lettuce grown in one standardised substrate under controlled conditions at five research centres. Knecht and O'Leary (1983) found no significant differences in the Cu content of lettuce grown with different levels of CO_2-enrichment, despite major effects of the latter on yield. The analytical values ranged from 3.4 to 5.1 μg g^{-1} Cu in the lettuce tops, and no deficiency symptoms were observed in any treatment.

35

Copper toxicity. High Cu contents are unfavourable to the growth of lettuce, an effect clearly shown in the response curves published by Moore *et al.* (1957); contents of up to 34 $\mu g\ g^{-1}$ Cu were recorded in the tops of the plants. MacKay *et al.* (1966) grew five successive crops of lettuce in a peat soil with six levels of added Cu. The fresh weight of plants containing 13 $\mu g\ g^{-1}$ Cu was 34% lower than that of plants containing 8 $\mu g\ g^{-1}$ Cu. A rather higher value of 21 $\mu g\ g^{-1}$ Cu was deduced as the critical level for toxicity in lettuce by Davis and Beckett (1978).

Zinc

Symptoms of deficiency. Zinc deficient plants have a rosette appearance and growth is restricted. At first there is no yellowing, but necrotic areas with a dark margin gradually develop near the edges of the leaves, particularly between the veins. The symptoms spread from the older leaves to the younger ones, but those parts of the leaves not exposed to the light remain green (Roorda van Eysinga and Smilde, 1971; see also the illustration reproduced by Scaife and Turner, 1983). In sand culture, plant size was reduced by Zn deficiency, followed by chlorosis of the leaves (Berry, 1971). Dull, dark purplish spots developed on the margins of mature leaves, accompanied by break-down of the laticifers. Later, dark interveinal lesions formed, coalesced and progressed towards the base of the leaves. The symptoms appeared on the older leaves and moved to the younger ones but rarely affected the very youngest leaves, even when the deficiency was severe.

Growth responses to zinc. Marlatt (1959, 1967) reported growth responses of field-grown lettuce to zinc sulphate, applied either as a foliar spray at approximately 900 mg l^{-1} or as a side dressing at 112 kg ha^{-1}. These responses were rarely accompanied by visual symptoms of Zn deficiency, however. Similar results were reported by Zink (1966). Thus application of zinc sulphate at rates of 11–38 kg ha^{-1} Zn increased the rates of growth and hence the yields at the first harvest. Weight per head was also increased following application yet no visual symptoms of deficiency were apparent in the untreated plots. Berry (1971) grew lettuce in solution culture at eight concentrations from 0.0016 to 0.2 mg l^{-1} Zn. The fresh and dry weights of the plants increased progressively with concentration up to 0.2 mg l^{-1}.

Zinc content of lettuce plants. Zink (1966) found growth responses to added Zn (see preceding paragraph) when plants in the standard (untreated) plots contained 63–79 $\mu g\ g^{-1}$ Zn. Such values would not normally be regarded as particularly low, and no deficiency symptoms were present. Plants from treated plots contained 94–108 $\mu g\ g^{-1}$ Zn. A critical level of 9 $\mu g\ g^{-1}$ Zn in mature mid-ribs of lettuce leaves, corresponding to a 10% reduction in growth, was deduced by Berry (1971). Somewhat higher concentrations were found in the laminae (with mid-ribs removed) of young and of mature leaves, the critical level for both being given as 17 $\mu g\ g^{-1}$ Zn. The overall ranges of Zn content were 6–31 $\mu g\ g^{-1}$ and 6–25 $\mu g\ g^{-1}$ for young and mature mid-rib tissue, with 11–61 $\mu g\ g^{-1}$ and 10–43 $\mu g\ g^{-1}$ in the young and mature laminae respectively. No significant differences in yield (fresh weight) were found over the range 0.05–0.2 mg l^{-1} Zn in the nutrient solutions, corresponding to 16–25 $\mu g\ g^{-1}$ Zn in the mature mid-ribs and 24–43 $\mu g\ g^{-1}$ in the mature laminae. The increase in fresh weight over this range amounted to 15%, however, giving some indication that the higher concentrations e.g. 25 and 43 $\mu g\ g^{-1}$ Zn in mature mid-ribs and laminae, might be beneficial. The Zn content of the roots was very similar to that of the mature laminae at all levels of supply and was about double that of the mature mid-ribs; a critical level of 18 $\mu g\ g^{-1}$ Zn was deduced for lettuce roots. Roorda van Eysinga and Smilde (1971) found 12 $\mu g\ g^{-1}$ Zn in deficient lettuce and 30–330 $\mu g\ g^{-1}$ in healthy plants. The same authors (1981) gave a value of 26 $\mu g\ g^{-1}$ Zn in the aerial parts of the plant, below which deficiency was likely to occur. Berry *et al.* (1981) found 39–71 $\mu g\ g^{-1}$ Zn in lettuce grown under controlled conditions in one standardised peat-vermiculite substrate at five different research centres.

Zinc toxicity. Marlatt (1959, 1967) found reduced yields of field-grown lettuce due to excessive application of chelated Zn (Zn-EDTA) as a foliar spray at about 2400 mg l^{-1} or as a side dressing of 224 kg ha^{-1}. The lower leaves first became chlorotic, then wilted and died. Similar symptoms were also recorded by Roorda van Eysinga and Smilde (1971). In addition, the latter authors sometimes found red-brown spots on the main veins and adjacent leaf tissue. In extreme cases the leaves were cupped and erect, and the plants failed to head. Particularly high levels (520–639 $\mu g\ g^{-1}$) of Zn were found in the leaves.

Localised zinc toxicity is sometimes encountered in glasshouses where condensate drips on to the plants from galvanised iron components such as the gutters. The symptoms are yellow on red/brown areas on the leaves (MAFF, 1982). The problem is increased where propane or paraffin are burned for CO_2 enrichment, possibly arising from the increased humidity or from chemical attack on the galvanising by oxides of sulphur formed during combustion.

Boron

Symptoms of deficiency. An early description of B deficiency in lettuce was given by McHargue and Calfee (1932). Growth in sand culture without added B was retarded after the first two weeks, and the newer leaves were pale in colour. In the fourth week, dark spots appeared on the tips of the young leaves, and these increased in size and spread along the leaf margins. The growing points died and turned black, followed eventually by death of the whole plant. Malformation of the leaves due to cessation of marginal growth was subsequently described and illustrated (McHargue and Calfee, 1933); see also accounts by Skinner (1941) and Purvis and Carolus (1964). Addition of B to plants initially deprived of this nutrient led to recovery; where the growing point had already died due to a deficiency, growth was resumed from axillary buds; see illustrations by McHargue and Calfee (1933).

Moinat (1943) described B deficient lettuce as having thick, brittle leaves, reduced in size and cupped in shape. Brown spots and waxy exudations appeared on the younger leaves, and the growing point gradually turned brown. The roots were short and stubby, with brown rather than white tips. Severe effects of deficiency in soil-grown lettuce were described by Midgley and Dunklee (1946). The responses included death of seedlings, die-back of transplants and yellowing of the inner (heart) leaves. The affected leaves later turned brown or black and frequently rotted. Death of the growing point was often followed by multiple branching, and viable seed was seldom produced. Messing (1954) observed only slight stunting and the development of tip-scorch on the younger leaves shortly before the plants reached maturity in sand culture.

Struckmeyer and Tibbitts (1965) observed that brownish spots first developed along the lower margins of the leaves. As these spots increased, the margins became scorched and the tissue between the veins was puckered. Anatomical studies showed collapse of cells some 3–4 layers from the periphery of the leaf. Laticifers in the affected areas were swollen and frequently burst, exuding latex both among the cells and to the outer surface of the leaf. Sometimes the release of latex stretched the epidermis, forming bulges visible externally as brown spots.

Roorda van Eysinga and Smilde (1971; 1981) noted that the leaves, especially the youngest ones, were stiff and rounded, with yellow mottling at the margins near the veins. The plants were darker than normal and severely stunted, with brown spotting of the very youngest leaves. Root development was extremely poor, and 'butterhead' types developed a rosette-like appearance. Puckering of the leaf margins and death of the growing point were illustrated by Scaife and Turner (1983).

Growth responses to boron. McHargue and Calfee (1933) studied the growth of lettuce at 21 concentrations of B in solution over the range of 0–5 mg l^{-1} B. The plants grew vigorously without obvious signs of deficiency within the range 0.4–0.9 mg l^{-1} B. Seed production was highest at 0.6–0.7 μg l^{-1} B, and an optimum of 0.7 μg l^{-1} was deduced. Moinat (1943) grew lettuce at ten concentrations of B (0–5 mg l^{-1} B) in sand culture. Even a very low

concentration (0.005 mg l^{-1} B) greatly increased growth, and no significant differences in yield (fresh or dry weight) were found between 0.005 and 2.5 mg l^{-1} B. Warington (1946) found a 65% reduction in growth when B was omitted in solution culture. Mackay (1962) grew lettuce in pots of peat at five levels of borax (0–40 lb/acre; 0–5.1 kg ha^{-1} B) in factorial combination with three treatments supplying lime and P. Maximum yields of dry matter were obtained with dressings of 20–40 lb borax per acre (2.5–5.1 kg ha $^{-1}$ B). Smilde (1972) studied the effects of liming on the response of lettuce in two types of peat. In a relatively unhumified sphagnum peat the lettuce responded positively to B applications at pH 5.1 and 5.8, but not at lower pH (3.3–4.1). In a well-humified old sphagnum peat ('black peat'), however, responses to B were found at pH 4.1 and above, though again not under highly acid conditions (pH 3.6). Adams *et al.* (1975) recorded a 25% depression of yield (fresh weight) in plots of peat from which B had been omitted.

Boron content of normal and deficient lettuce. Deficient plants contain less than 20 μg g^{-1} B and healthy ones 25–50 μg g^{-1} (McHargue and Calfee, 1933). Eaton (1945) recorded 27–43 μg g^{-1} B in the leaves of lettuce without deficiency symptoms in sand culture. Midgley and Dunklee (1946) found 14–28 μg g^{-1} B in the yellow leaves of boron-deficient lettuce, whereas healthy leaves contained 35 μg g^{-1} B. Mackay *et al.* (1962) plotted yield and content of the plants against rate of borax application for an acid peat; a 'minimum sufficiency' concentration of 23 μg g^{-1} B in the dry matter (excluding roots) was deduced.

Marlatt (1967) quoted 35 μg g^{-1} B in healthy lettuce and 12–28 μg g^{-1} B in deficient leaves; these values are very close to those given by Midgley and Dunklee (1946). The contents of different parts of lettuce leaves recorded by Ashkar and Ries (1971) ranged from 39 μg g^{-1} B in the leaf margins to 24 μg g^{-1} in the midrib and 25 μg g^{-1} in the remainder of the leaf. Roorda van Eysinga and Smilde (1971) found relatively low values of 6–10 μg g^{-1} B in the heads of B deficient lettuce compared with 32–37 μg g^{-1} in normal plants. Smilde (1972) concluded that responses to added B were normally associated with leaf concentrations below 16 μg g^{-1} B. Values of 16–29 μg g^{-1} B were recorded by Berry *et al.* (1981) in a collaborative study at five centres, each using the same peat-vermiculite mixture with added nutrients under controlled environmental conditions.

Boron toxicity. Lettuces are easily damaged by excess B and the margin between sufficiency and toxicity is small. Thus McHargue and Calfee (1933) reported slight chlorosis due to excess at a concentration of only 0.9 mg l^{-1} B in the culture solution, this being only 0.2 mg l^{-1} above the optimum for plant growth. Symptoms observed at 1.2 mg l^{-1} B included chlorosis and death of the older leaves and large, white, dead areas on the edges of mature living leaves. The dry weight of the plants decreased significantly at and above 1.2 mg l^{-1} B, and concentrations above 2.5 mg l^{-1} proved fatal to the young plants. Toxicity was associated with concentrations above 60 μg g^{-1} B in the leaves.

Moinat (1943) reported tissue death and the formation of large brown bands along the margins of the older leaves when growing in sand culture at 2.5 and 5.0 mg l^{-1} B. Eaton (1945) recorded a mild marginal leaf scorch in lettuce grown at 1 mg l^{-1} B, corresponding to 70 μg g^{-1} B in the leaves. More severe leaf scorch occurred at still higher concentrations in the nutrient solution (5, 10, 15 and 25 mg l^{-1} B), with concentrations of up to 582 μg g^{-1} B in the leaves. In his review of nutritional disorders of lettuce, Marlatt (1967) referred to chlorosis of the lower leaves, with large white or tan dead spots, especially along the margins. Roorda van Eysinga and Smilde (1971, 1981) described and illustrated toxicity, including sunken, brownish-grey spots developing on the margins of the older leaves. As the toxicity became more severe, the spots coalesced and dried out. The symptoms progressed towards the younger leaves, though the very youngest leaves appeared to be unaffected at first. The leaf B contents associated with toxicity ranged from 40 μg g^{-1} to 200 μg g^{-1}, depending on severity.

Molybdenum

Symptoms of deficiency. Lettuce grown in solution culture without added Mo were small and pale, with a loose, open habit (Warington, 1946). A frill of withered leaves later developed between the still healthy inner and outer whorls. The withering started as patches of crinkled tissue, mainly at the tips of the intermediate leaves; the patches coalesced and turned yellow and papery. In less severe cases only colour and size were affected, without specific symptoms. Plants grown in sand culture (Hewitt and Bolle-Jones, 1952) developed small, pale yellow-green leaves, these later becoming yellow and inrolled, with scorched margins. The symptoms commenced in the oldest leaves and progressed to the youngest, and the plants withered and died within 30–35 days. Von Stieglitz and Chippendale (1955) described the older leaves of deficient lettuce as having a dull yellow or whitish appearance, often more intense near the edges. Death of interveinal tissue was also noted in severe cases.

Rubins (1955) recorded that lettuce grown in two soils without added Mo were light green in colour. Dead, light brown areas of tissue developed along the margins of the leaves and, in some cases, the plants died. Plant (1956) referred to Mo deficient lettuce as chlorotic and stunted in the early stages. The laminae failed to develop and the leaves were ovate rather than round, with marginal chlorosis. The leaves became 'papery' and almost translucent. In severe cases the plants failed to heart and

eventually died. Marlatt (1967) referred to yellowing and marginal necrosis of the outer leaves. Interveinal areas sometimes became chlorotic and died. The plants were stunted and hearting was delayed or prevented.

Roorda van Eysinga and Smilde (1971) referred to a widespread occurrence of wilting leaves amongst young lettuce plants being raised in peat blocks during 1968. The cause was shown to be Mo deficiency. The leaves of the young plants wither from the tip and margins inward. Transparent spots, which later become brown and coalesce, may develop on older plants. Growth of both roots and foliage is severely stunted. According to van den Ende and Boertje (1971), the older leaves first turn pale green and later yellow, thought the cotyledons retain almost their normal colour. The leaves flag and, beginning at the tips, may shrivel and die.

Growth responses to molybdenum. Beneficial effects on the growth of lettuce in solution culture were reported by Brenchley and Warington (1942). Plants receiving Mo at 0.1 mg l^{-1} were larger, deeper green and apparently more resistant to disease. Omission depressed the dry weight of two lettuce cultivars by 77–87% (Warington, 1946). Responses of field-grown lettuce to Mo were reported by Wilson (1948) in Australia and by Plant (1952, 1956) in the U.K. Thus Plant (1952) recorded yield increases of 41% and 100% from applications of sodium molybdate at 2.2 and 4.5 kg ha^{-1} to a soil of pH 5.2 known to cause Mo deficiency in cauliflower. Rubins (1955) found yield reductions (dry weight) of 46–64% where Mo was not applied to three acid soils (pH 4.9–5.2). Marlatt (1967) noted that application of sodium molybdate at 2.2–4.5 kg ha^{-1} would often eliminate the deficiency; this salt could also be applied as a foliar spray (0.07 g per plant in 50–100 ml water). Van den Ende and Boertje (1971) eliminated deficiency symptoms and increased yields of lettuce by application of ammonium molybdate to a peat substrate at a rate of 6 g m^{-3}, and recommended a dressing of 5 g m^{-3}. Deficiency symptoms in young plants could be controlled by spraying with a 0.05–0.1% solution of ammonium molybdate. In cases of acute deficiency Roorda van Eysinga and Smilde (1971) recommended applying ammonium or sodium molybdate at 0.15 g m^{-2} in ample water, or spraying with these salts at 0.05%. Gupta et al. (1978) reported yield increases (dry matter) of 60% from pre-treatment of lettuce seed with molybdate and of 88% where ammonium molybdate was applied to the peat substrate (mean for five rates of application).

Warington (1950) found greater yield responses to Mo at normal than at low levels of NO$_3$-N. Thus, whilst no positive response to Mo (0.1 mg l^{-1}) resulted at low N (35 mg l^{-1}), yield increases of 27% and 93% were obtained at 70 mg l^{-1} N and 139 mg l^{-1} N respectively. Similarly, in beds of peat, there was little response to Mo at the lower levels of N (nitro-chalk) applied but an 8% yield increase at the highest level of N (Adams et al., 1975). Additional

N in the form of sulphur-coated urea depressed the yields by about 20% but increased the responses to Mo . Thus in peat plots (pH 5.4) receiving both nitro-chalk and sulphur-coated urea, added Mo increased the yield by 15% and raised the proportion of hearted and marketable lettuce by 67% and 17% respectively.

Molybdenum supply and nitrate content of lettuce. Since molybdenum is a constituent of the enzyme nitrate reductase, there is a tendency for NO_3-N to accumulate in Mo deficient plants when supplied with N in this form. Thus Warington (1950) recorded 0.22% NO_3-N in the dry matter of lettuce grown without added Mo at an NO_3-N concentration of 139 mg l^{-1} compared with 0.04% and 0.02% NO_3-N in the plants where Mo was supplied at 0.1 and 5.0 mg l^{-1}. Lettuce grown in three Mo deficient soils (Rubins, 1955) had NO_3-N levels of 0.21–0.33% NO_3-N in the heads, decreasing by 72–87% where Mo was applied. A far higher level of 1.17% NO_3-N in the dry matter was found in young lettuce grown in an acid peat (pH 4.0), decreasing to 0.39% NO_3-N where Mo was supplied; the total N content was unchanged (van den Ende and Boertje, 1971). Concentrations of 42–48 mg l^{-1} NO_3-N were found in the leaf sap from plants without Mo compared with 12–14 mg l^{-1} NO_3-N where Mo was applied to the peat substrate.

Effects of pH. The availability of Mo to plants is very dependent on the pH of the substrate. For example, Plant (1952, 1956) reported a yield increase of 84% due to liming an acid soil from pH 5.2 to pH 6.3, accompanied by some small increase in the Mo content of the leaves. For comparison, the yield responses to added sodium molybdate in this same soil, without added lime, were 41% and 100% at rates of 2.24 and 4.48 kg ha^{-1}. Sorteberg (1954) found that healthy lettuce plants could be grown on an acid substrate either by applying ammonium molybdate (1 kg ha^{-1}) or by moderate liming to pH 5.9. Rubins (1955) reported greater yield responses from liming acid soils (pH 4.7–5.9) than from addition of sodium molybdate. It was concluded that the soils used in these experiments had adequate reserves of Mo which were readily released by liming. Von Stieglitz and Chippendale (1955) noted that Mo deficiency in lettuce was particularly associated with strongly acid soils. In a peat substrate, liming from pH 4.4 to 6.5 without added Mo increased yields by 587%, accompanied by a decrease in visual symptoms of deficiency (van den Ende and Boertje, 1971). The responses to added Mo were also greatest under acid conditions, the increases in weight of young lettuce being 760% at pH 4.4, 59% at pH 5.5 and only 11% at pH 6.5: similar results were obtained by Smilde (1972). Thus, interpolated from the graphical data, the maximum increases in lettuce yield when sodium molybdate was applied to a moss peat were 267% at pH 4.1, 160% at pH 5.1 and 37% at pH 5.8.

Again, liming greatly increased yields even without addition of molybdate. Adams *et al.* (1975) found that omission of Mo from a peat substrate caused severe stunting and some death of the plants at pH 5.0, whereas growth was only slightly affected at pH 6.5.

Molybdenum content of deficient and normal lettuce. Lettuce grown in solution culture (139 mg l^{-1} NO_3-N) with and without Mo (0.1 mg l^{-1} Mo) contained 0.90 and 0.26 µg g^{-1} Mo respectively in the dry matter; a small increase in yield (9%) resulted from inclusion of Mo in the solution (Warington, 1950). Considerably lower contents of 0.03–0.04 µg g^{-1} Mo were recorded in further experiments where this element was not supplied: addition increased the yields by 16–35% in three out of four instances, the treated plants containing 0.20–0.69 µg g^{-1} Mo. Plant (1952, 1956) reported levels of 0.06 and 0.08 µg g^{-1} Mo in deficient and healthy lettuce, accompanied by a doubling of yield; both values seem quite low, though, as noted by Plant, the total uptake of Mo by the crop was considerably increased by Mo application. MacKay *et al.* (1966) reported rather higher values, ranging from 0.5 µg g^{-1} Mo in lettuce grown without added Mo to 3.2 µg g^{-1} at the two most favourable rates of application. Marlatt (1967) concluded that deficient leaves may contain 0.06 µg g^{-1} Mo whereas normal leaves contain about 0.1 µg g^{-1} Mo.

Roorda van Eysinga and Smilde (1971) reported 0.3 µg g^{-1} Mo in young lettuce showing distinct deficiency symptoms compared with 2.3–3.7 µg g^{-1} Mo in healthy plants. Van den Ende and Boertje (1971) grew lettuce in a high-moor ('white') peat with and without Mo; deficient leaves contained 0.22 µg g^{-1} Mo compared with 0.5 µg g^{-1} in healthy leaves. Molybdenum concentrations in lettuce (tops) grown in two contrasting types of peat with four levels each of sodium molybdate and lime in factorial combination were reported by Smilde (1972). It was concluded that responses to Mo were normally associated with leaf concentrations below 0.12–0.19 µg g^{-1} Mo. Further examination of the published data suggests that the Mo content of leaves corresponding to maximum growth increases with the pH of the substrate and, for a given pH level, is higher in the white than in the black (humified) peat. Pre-treatment of lettuce seed with Mo increased the yield in an acid peat soil by 60% without raising the content of the leaves (0.2 µg g^{-1} Mo). Molybdenum added to the soil at a rate of 4.5 µg g^{-1} increased the yield by 90% and raised the leaf content to 0.4 µg g^{-1} Mo; little difference in yield was found at five rates of application corresponding to 0.4–9.0 µg g^{-1} Mo in the leaves. Berry *et al.* (1981) found 1.2–2.7 µg g^{-1} Mo in the tops of lettuces grown under controlled conditions at five collaborating research centres.

The rather wide range of values reported in this section may be associated partly with problems of determining low concentrations of Mo in plant tissue.

Molybdenum toxicity in lettuce. Warington (1946) grew lettuce at a wide range of Mo concentrations (0–100 mg l^{-1}) in solution culture. Growth was remarkably uniform over the range 0.001–10 mg l^{-1} Mo. Toxic effects were invariably present at 100 mg l^{-1} Mo, though their severity varied with variety and time of year. The first symptom of toxicity was a yellow-brown discoloration of the roots; growth was much reduced and the leaves turned a golden yellow colour. A similar golden colour has been noted in other plants affected by excess Mo, and is associated with the formation of globular yellow bodies within the cells, apparently a tannin-molybdenum complex (Warington, 1937, 1954). Slight root discolouration occurred even at 10 mg l^{-1} Mo in one variety ('Tom Thumb'; Warington, 1946). Yield reductions at 10 mg l^{-1} Mo were later reported by Warington (1950), ranging from 9% at low N (35 mg l^{-1} N) to 15% and 26% at 70 mg l^{-1} and 139 mg l^{-1} N respectively; lettuce grown at 10 mg l^{-1} Mo in solution contained 65–73 $\mu g\ g^{-1}$ Mo in their tops. MacKay *et al.* (1966) noted a wide range of tolerance to excess Mo in soil-grown lettuce. Even at the highest level of application, corresponding to 277 $\mu g\ g^{-1}$ in the plants, the yield (averaged over five crops) was only reduced by 6.3%.

References

Adams, P., Graves, C. J. and Winsor, G. W. (1975). Nutrition of lettuce in peat. *Rep Glasshouse Crops Res Inst 1974*, 70–71.

Adams, P., Graves, C. J. and Winsor, G. W. (1978a). Some responses of lettuce, grown in beds of peat, to nitrogen, potassium, magnesium and molybdenum. *J hort Sci* **53**, 275–281.

Adams, P., Graves, C. J. and Winsor, G. W. (1978b). Effects of copper deficiency and liming on the yield, quality and copper status of tomatoes, lettuce and cucumbers grown in peat. *Scientia Hort* **9**, 199–205.

Adams, P. and Winsor, G. W. (1973). Analysis of soil solution as a guide to the nutrient status of glasshouse soils. *Pl Soil* **39**, 649–659.

Adams, P. and Winsor, G. W. (1984). Some responses of cucumbers and lettuce, grown in a peat substrate, to phosphate and lime. *Acta hort* **145**, 102–111.

Anon (1964). Lettuce: The effect of four different sources of nitrogen. *Rep Fairfield exp Hort Stn 1963*, 31–32.

Ashkar, S. A. and Ries, S. K. (1971). Lettuce tipburn as related to nutrient imbalance and nitrogen composition. *J Am Soc hort Sci* **96**, 448–452.

Berry, W. L. (1971). The nutrient status of zinc in lettuce evaluated by plant analysis. *J Am Soc hort Sci* **96**, 412–414.

Berry, W. L. and Carey, R. (1971). Evaluation of the potassium nutrient status of seedling lettuce by plant analysis. *J Am Soc hort Sci* **96**, 298–300.

Berry, W. L., Krizek, D. T., Ormrod, D. P., McFarlane, J. C., Langhans, R. W. and Tibbitts, T. W. (1981). Variation in elemental content of lettuce grown under base-line conditions in five controlled-environment facilities. *J Am Soc hort Sci* **106**, 661–666.

Bishop, R. F., Chipman, E. W. and MacEachern, C. R. (1973). Effect of nitrogen, phosphorus and potassium on yields and nutrient levels in celery and head lettuce grown on sphagnum peat. *Commun Soil Sci Pl Anal* **4**, 375–387.

Borkowski, J. (1977). Effect of calcium, nitrogen and potassium fertilisation on the occurrence of physiological disorders and on growth and yield of lettuce. *Biuletyn Warzywniczy* **20**, 345–349.

Borkowski, J., Szwonek, E. and Ostrzycka, J. (1975). The effect of different levels of fertilisation with potassium on the growth of lettuce in greenhouses, its commercial value and the potassium and magnesium status. *Roczn Nauk roln* **100**, 139–151.

Brenchley, W. E. and Warington, K. (1942). Value of molybdenum for lettuce. *Nature Lond* **149**, 196.

Cantliffe, D. J. and Phatak, S. C. (1974). Nitrate accumulation in greenhouse vegetable crops. *Can J Pl Sci* **54**, 783–788.

Collier, G. F. and Holness, P. (1984). Crop quality: Lettuce tipburn: Tipburn induction *Rep natn Veg Res Stn 1983*, 123.

Collier, G. F. and Huntington, V. C. (1983). The relationship between leaf growth, calcium accumulation and distribution, and tipburn development in field-grown butterhead lettuce. *Scientia Hort* **21**, 123–128.

Collier, G. F. and Tibbitts, T. W. (1984). Effects of relative humidity and root temperature on calcium concentration and tipburn development in lettuce. *J Am Soc hort Sci* **109**, 128–131.

Davidson, J. H. and Thiegs, B. J. (1966). Fumigation side effect. *Biokemia No 13*, 15–17.

Davies, J. N. (1957). Steam sterilisation studies. *Rep Glasshouse Crops Res Inst 1954–1955*, 70–79.

Davis, R. D. and Beckett, P. H. T. (1978). The use of young plants to detect metal accumulations in soils. *Wat Pollut Control* **77**, 193–205.

Eaton, F. M. (1945). Deficiency, toxicity, and accumulation of boron in plants. *J agric Res* **69**, 237–277.

Ende, J. van den and Boertje, G. A. (1971). Molybdenum deficiency in young lettuce and tomato plants. *Acta Hort* **26**, 61–67.

Frota, J. N. E. and Tucker, T. C. (1972). Temperature influence on ammonium and nitrate absorption by lettuce. *Proc Soil Sci Soc Am* **36**, 97–100.

Gardner, B. R. and Pew, W. D. (1972). Response of fall grown head lettuce to nitrogen fertilisation. *Ariz Agric Exp Stn Tech Bull* **199**, 8pp.

Gardner, B. R. and Pew, W. D. (1974). Response of spring grown head lettuce to nitrogen fertiliser. *Ariz Agric Exp Stn Tech Bull* **210**, 9pp.

Gilbert, B. E. and McLean, F. T. (1928). A 'deficiency disease': The lack of available manganese in a lime-induced chlorosis. *Soil Sci* **26**, 27–31.

Goodall, D. W., Grant Lipp, A. E. and Slater, W. G. (1955). Nutrient interactions and deficiency diagnosis in the lettuce. I. Nutritional interaction and growth. *Aust J Biol Scis* **8**, 301–329.

Grant Lipp, A. E. and Goodall, D. W. (1958a). Nutrient interactions and deficiency diagnosis in the lettuce. IV. Phosphorus content and response to phosphorus. *Aust J Biol Scis* **11**, 30–44.

Grant Lipp, A. E. and Goodall, D. W. (1958b). Nutrient interactions and deficiency diagnosis in the lettuce. V. Potassium content and response to potassium. *Aust J Biol Scis* **11**, 471–484.

Gupta, U. C., Chipman, E. W. and MacKay, D. C. (1978). Effects of molybdenum and lime on the yield and molybdenum concentration of crops grown on acid sphagnum peat soil. *Can J Pl Sci* **58**, 983–992.

Harward, M. E., Jackson, W. A., Lott, W. L. and Mason, D. D. (1955). Effects of Al, Fe and Mn upon the growth and composition of lettuce. *Proc Am Soc hort Sci* **66**, 261–266.

Haworth, F. and Cleaver, T. J. (1967). The effects of different manurial treatments on the yield and mineral composition of winter lettuce. *J hort Sci* **42**, 23–29.

Hewitt, E. J. and Bolle-Jones, E. W. (1952). Molybdenum as a plant nutrient. II. The effects of molybdenum deficiency on some horticultural and agricultural crop plants in sand culture. *J hort Sci* **27**, 257–265.

Ikeda, H. and Osawa, T. (1980). Comparison of adaptability to nitrogen source among vegetable crops. II. Growth response and accumulation of ammonium and nitrate-nitrogen of leaf vegetables cultured in nutrient solution containing nitrate, ammonium and nitrite as nitrogen sources. *J Jap Soc hort Sci* **48**, 435–442.

Jager, G., van der Boon, J. and Rauw, G. J. G. (1969a). The influence of soil steaming on some properties of the soil and on the growth and heading of winter glasshouse lettuce. I. Changes in chemical and physical properties. *Neth J agric Sci* **17**, 143–152.

Jager, G., van der Boon, J. and Rauw, G. J. G. (1969b). The influence of soil steaming on some properties of the soil and on the growth and heading of winter glasshouse lettuce. II. The reaction of the crop. *Neth J agric Sci* **17**, 241–245.

Jager, G., van der Boon, J. and Rauw, G. J. G. (1970). The influence of soil steaming on some properties of the soil and on the growth and heading of winter lettuce. III. The influence of nitrogen form, manganese level and shading studied in sand culture experiments with trickle irrigation. *Neth J agric Sci* **18**, 158–167.

Knavel, D. E. (1974). The influence of growing temperatures and leaf trimming on nutrient content in lettuce. *HortSci* **9**, 231–232.

Knavel, D. E. (1981). The influence of temperature and nutrition on the growth and composition of lettuce. *Hort Res* **21**, 11–18.

Knecht, G. N. and O'Leary, J. W. (1983). The influence of carbon dioxide on the growth, pigment, protein, carbohydrate and mineral status of lettuce. *J Pl Nutr* **6**, 301–312.

Krauss, A. (1971). [The effect of fertilising lettuce with the major nutrients on attack by *Botrytis cinerea*]. *Z PflErnähr Bodenk* **128**, 12–23; *Hort Abstr* **41**, No. 6394.

Lambeth, V. N. (1953). Variable potassium and magnesium saturation on growth and mineral composition of Bibb lettuce. *Proc Am Soc hort Sci* **62**, 357–362.

Lauder, K. (1977). Lettuce on concrete. *Grower* **87** (Feb 24 suppt), 40, 41, 44–46.

Lorenz, O. A. and Minges, P. A. (1942). Nutrient absorption by a summer crop of lettuce in Salinas Valley, California. *Proc Am Soc hort Sci* **40**, 523–527.

Lorenz, O. A., Tyler, K. B. and McCoy, O. D. (1964). Phosphate sources and rates for winter lettuce on a calcareous soil. *Proc Am Soc hort Sci* **84**, 348–355.

MacKay, D. C., Langille, W. M and Chipman, E. W. (1962). Boron deficiency and toxicity in crops grown on sphagnum peat soil. *Can J Soil Sci* **42**, 302–310.

MacKay, D. C. and Chipman, E. W. (1961). The response of several vegetables to applied nitrogen, phosphorus and potassium on a sphagnum peat soil. *Proc Soil Sci Soc Am* **25**, 309–312.

MacKay, D. C., Chipman, E. W. and Gupta, U. C. (1966). Copper and molybdenum nutrition of crops grown on acid sphagnum peat soil. *Proc Soil Sci Soc Am* **30**, 755–759.

Marlatt, R. B. (1959). The effects of applications of micronutrients to Arizona field-grown lettuce. *Pl Dis Reptr* **43**, 1019–1022.

Marlatt, R. B. (1967). Nonpathogenic diseases of lettuce, their identification and control. *Tech Bull* **721**. *Fla agric exp Stn*, 39pp.

Massey, D. M. (1975). Nutrition and environment. *Rep Glasshouse Crops Res Inst 1974*, 73.

Massey, D. M. and Winsor, G. W. (1960). Manganese toxicity. *Rep Glasshouse Crops Res Inst 1959*, 37–39.

Massey, D. M. and Winsor, G. W. (1974). Nutrition and environment. *Rep Glasshouse Crops Res Inst 1973*, 74–75.

McHargue, J. S. and Calfee, R. K. (1932). Effect of boron on the growth of lettuce. *Pl Physiol* **7**, 161–164.

McHargue, J. S. and Calfee, R. K. (1933). Further evidence that boron is essential for the growth of lettuce. *Pl Physiol* **8**, 305–313.

Messing, J. H. L. (1954). The effect of some acute mineral deficiencies on plant growth. (e) Lettuce. *Rep exp Res Stn Cheshunt 1953*, 57.

Messing, J. H. L. (1965a). The effects of lime and superphosphate on manganese toxicity in steam-sterilized soil. *Pl Soil* **23**, 1–16.

Messing, J. H. L. (1965b). Some differences in the growth of lettuce varieties at a high manganese level in sand culture. *Rep Glasshouse Crops Res Inst 1964*, 142–148.

Messing, J. H. L. (1971). Metal toxicities in glasshouse crops. *Tech Bull Minist Agric Fish Fd* **21**, 159–175.

Midgley, A. R. and Dunklee, D. E. (1946). Boron deficiency of lettuce. *Better Crops Pl Fd* **30**, No. 2, 17–20, 42–43.

Ministry of Agriculture, Fisheries and Food (1982). Protected lettuce production 2. Soil preparation, nutrition and planting. Leaflet **662**, 9pp.

Moinat, A. D. (1943). Nutritional relationships of boron and indoleacetic acid in head lettuce. *Pl Physiol* **18**, 517–523.

Mondrzak-Rosenberg, G. (1965). The effect of nitrogen and phosphorus on the yield of Romaine lettuce. *Israel J agric Res* **15**, 209–210.

Moore, D. P., Harward, M. E., Mason, D. D., Hader, R. J., Lott, W. L. and Jackson, W. A. (1957). An investigation of some of the relationships between copper, iron and molybdenum in the growth and nutrition of lettuce: II. Response surfaces of growth and accumulation of copper and iron. *Proc Soil Sci Soc Am* **21**, 65–74.

Nicholas, D. J. D. (1949). The manganese and iron contents of crop plants as determined by chemical methods. *J hort Sci* **25**, 60–77.

Pandita, M. L. and Andrew, W. T. (1967). A correlation between phosphorus content of leaf tissue and days to maturity in tomato and lettuce. *Proc Am Soc hort Sci* **91**, 544–549.

Parups. E. V. and Goodwin-Wilson, R. (1958). Nutrition of lettuce with nitrogen, phosphorus and potassium on organic soils in Ontario. *Proc Am Soc hort Sci* **71**, 399–406.

Paul, J. L. and Polle, E. (1965). Nitrite accumulation related to lettuce growth in a slightly alkaline soil. *Soil Sci* **100**, 292–297.

Perez Melian, G., Luque Escalona, A. and Steiner, A. A. (1977). Leaf analysis as a diagnosis of nutritional deficiency or excess in the soilless culture of lettuce. *Pl Soil* **48**, 259–267.

Plant, W. (1952). Molybdenum deficiency in lettuce. *Nature Lond* **169**, 803.

Plant, W. (1956). The effect of molybdenum deficiency and mineral toxicity on crops in acid soils. *J hort Sci* **31**, 163–176.

Purvis, E. R. and Carolus, R. L. (1964). In H. B. Sprague, ed. Hunger Signs in Crops, 3rd ed. 245–286. *David McKay Co NY*.

Roorda van Eysinga, J. P. N. L. (1966). Fertilisation of lettuce grown in glasshouses with dried blood and nitrochalk. *Publties Proefstn Groenten-en Fruitteelt Glas Naaldwijk* **110**, 18.

Roorda van Eysinga, J. P. N. L. (1984). Nitrate and glasshouse vegetables. *Fert Res* **5**, 149–156.

Roorda van Eysinga, J. P. N. L. and Smilde, K. W. (1971). Nutritional disorders in glasshouse lettuce. *Cent agric Publ Documn, Wageningen*, 56pp.

Roorda van Eysinga, J. P. N. L. and Smilde, K. W. (1981). Nutritional disorders in glasshouse tomatoes, cucumbers and lettuce. *Cent agric Publ Documn, Wageningen*, 130pp.

Rubins, E. J. (1955). Molybdenum status of some uncultivated Connecticut soils. *Proc Soil Sci Soc Am* **19**, 207–209.

Scaife, A. and Turner, M. (1983). J. B. D. Robinson, ed. Diagnosis of Mineral Disorders in Plants. *Vol 2*, 96pp. *HMSO Lond*.

Schmehl, W. R., Peech, M. and Bradfield, R. (1950). Causes of poor growth of plants on acid soils and beneficial effects of liming: I. Evaluation of factors responsible for acid-soil injury. *Soil Sci* **70**, 393–410.

Skinner, J. J. (1941). *In* G. Hambidge, ed. Hunger signs in crops, 149–189. Washington *Am Soc Agron & Nat Fert Assoc.*

Slater, W. G. and Goodall, D. W. (1957). Nutrient interactions and deficiency diagnosis in the lettuce. III. Nitrogen content and response to nitrogen. *Aust J Biol Scis* **10**, 253–278.

Smith, R. and Scaife, M. A. (1973). The phosphorus requirement of lettuce. I. Use of P intensity estimates to predict the response curve. *J agric Sci Camb* **80**, 111–117.

Smilde, K. W. (1972). Trace nutrient requirements of some plant species on peat substrates. *Proc 4th Int Peat Congr Helsinki* **3**, 239–259.

Sonneveld, C. (1968). De mangaanhuishouding van de grond en de mangaanopname van sla. *Meded Dir Tuinb* **31**, 476–483.

Sonneveld, C. and Voogt, S. J. (1975a). Studies on the manganese uptake of lettuce on steam-sterilised glasshouse soils. *Pl Soil* **42**, 49–64.

Sonneveld, C. and Voogt, S. J. (1975b). The effect of steam sterilisation of soil on the manganese uptake by glasshouse crops. *Act Hort* **51**, 311–319.

Sorteberg, A. (1954). Fortsatte forsøk med molybden. [Continued experiments with molybdenum]. *Forskn Fors Landbr* **5**, 161–198; *Soils Fertil* **17**, 438 (Abstr 2038).

Von Stieglitz, C. R. and Chippendale, F. (1955). Nutritional disorders of plants. *Qd. Dep Agric Stk.* 36pp.

Struckmeyer, B. E. and Tibbitts, T. W. (1965). Anatomy of lettuce leaves grown with a complete nutrient supply and without calcium or boron. *Proc Am Soc hort Sci* **87**, 324–329.

Thibodeau, P. O. and Minotti, P. L. (1969). The influence of calcium on the development of lettuce tipburn. *J Am Soc hort Sci* **94**, 372–376.

Tibbitts, T. W., Struckmeyer, B. E. and Rama Rao, R. (1965). Tipburn of lettuce as related to release of latex. *Proc Am Soc hort Sci* **86**, 462–467.

Twyman, E. S. (1951). The iron and manganese requirements of plants. *New Phytol* **50**, 210–226.

Tyler, K. B. and Lorenz, O. A. (1962). Diagnosing nutrient needs in vegetables. *Better Crops with Pl Fd* **46** (3) 6–13.

Vlamis, J. and Williams, D.E. (1973). Manganese toxicity and marginal chlorosis of lettuce. *Pl Soil* **39**, 245–251.

Wallace, T. (1951). The Diagnosis of Mineral Deficiencies in Plants by Visual Symptoms. 2nd ed. 107pp. *HMSO Lond.*

Wallace, T. (1961). The Diagnosis of Mineral Deficiencies in Plants by Visual Symptoms. 3rd Ed 125pp. *HMSO Lond.*

Warington, K. (1937). Observations on the effect of molybdenum on plants with special reference to the Solanaceae. *Ann appl Biol* **24**, 475–493.

Warington, K. (1946). Molybdenum as a factor in the nutrition of lettuce. *Ann appl Biol* **33**, 249–254.

Warington, K. (1950). The effect of variations in calcium supply, pH value and nitrogen content of nutrient solutions on the response of lettuce and red clover to molybdenum. *Ann appl Biol* **37**, 607–623.

Warington, K. (1954). Recent work on molybdenum and some micro-nutrient interactions. *Rep Rothamsted exp Stn 1953*, 181–187.

Williams, D. E. and Vlamis, J. (1957). Manganese toxicity in standard culture solutions. *Pl Soil* **8**, 183–193.

Wilson, R. D. (1948). Some responses of lettuce to the application of molybdenum. *J Aust Inst agric Sci* **14**, 180–187.

Winsor, G. W. (1968). Potassium and the quality of glasshouse crops. *Proc 8th Cong Int Pot Inst Brussels 1966*, 303–312.

Winsor, G. W. (1969). Nitrogen and glasshouse crops. *Minist Agric Fish Fd Tech Bull* **15**, 109–128.

Winsor, G. W. and Adams, P. (1968a). The nutrition of glasshouse crops. Lettuce nutrition: 3^2 N K × 2^3 P Mg Ca factorial study. *Rep glasshouse Crops Res Inst 1967*, 65–66.

Winsor, G. W. and Adams, P. (1968b). The nutrition of glasshouse crops. Lettuce nutrition: Responses to phosphate and lime. *Rep Glasshouse Crops Res Inst 1967*, 63–64.

Winsor, G. W., Davies, J. N. and Long, M. I. E. (1967). The effects of nitrogen, phosphorus, potassium, magnesium and lime in factorial combination on the yield of glasshouse tomatoes. *J hort Sci* **42**, 277–288.

Winsor, G. W. and Long, M. I. E. (1961). Nutritional trials with glasshouse crops: 3^2 × 2^3 factorial nutritional trials with lettuce. *Rep Glasshouse Crops Res Inst 1960*, 50–54.

Winsor, G. W. and Long, M. I. E. (1963a). Nutritional trials with glasshouse and other crops. 3^2 × 2^3 factorial nutritional trial with winter lettuce. *Rep Glasshouse Crops Res Inst 1962*, 63–64.

Winsor, G. W. and Long, M. I. E. (1963b). Some effects of potassium and lime on the relation between phosphorus in soil and plant, with particular reference to glasshouse tomatoes, carnations and winter lettuce. *J Sci Fd Agric* **14**, 251–259.

Winsor, G. W., Long, M. I. E. and Hart, B. (1964). Nutritional trials with glasshouse and other crops. 3^2 × 2^3 factorial nutritional trial with winter lettuce. *Rep Glasshouse Crops Res Inst 1963*, 69–70.

Woodman, R. M. (1939). Studies in the nutrition of vegetables. The effects of variation in the nitrogen supply on lettuce (var. May King) in sand culture. *Ann Bot Lond NS* **3**, 649–656.

Zink, F. W. (1966). The response of head lettuce to soil application of zinc. *Proc Am Soc hort Sci* **89**, 406–414.

Zink, F. W. and Yamaguchi, M. (1962). Studies on the growth rate and nutrient absorption of head lettuce. *Hilgardia* **32**, 471–500.

Pepper

The sweet pepper *(Capsicum annuum L.)* has a similar geographical origin to the tomato, namely Central and South America, and grows well under similar conditions to the tomato under glass. Both crops are rich in vitamin C. Peppers crop well with minimum temperatures of 20° and 18°C during the day and night respectively. Lower night temperatures delay picking and reduce the early yield, but have little effect on the cummulative yield after 8–10 weeks harvesting (Hand, 1980). General notes on the cultivation of sweet peppers have been published by the Ministry of Agriculture (MAFF, 1981).

Nitrogen

Symptoms and growth responses. Plants with N deficiency showed weak, stunted growth; the leaves were yellow and the fruit was yellow-green in colour (Miller, 1961). Plants grown at a low N level in sandy soil had thinner, less branched stems than those receiving fully adequate N. Leaves formed during the later stages of growth were small and light yellowish green, with short petioles (Cochran, 1936).

In sand culture, flower production of an early crop increased at first with the level of N applied, but this trend was later reversed and the overall response was not significant. A low level of N always reduced the number of fruit formed whilst a high level generally increased the number of fruit and the proportion of flowers that set (Maynard *et al.*, 1962). These results are generally similar to those reported earlier (Cochran, 1936). Responses to N have been somewhat variable (Pimpini, 1967), depending on soil type and, for field crops, on the rainfall. For example, in one year increasing N levels depressed the yield but had no effect on plant height, whereas an increase in both height and yield were found in the following year (Singh and Nettles, 1962). In general, the yield was depressed by N added to soil with an adequate content (Bottini, 1968). Thomas and Heilman (1964) reported a progressive increase in the yield of field-grown peppers in Texas over the range 0–135 kg ha^{-1} N, but higher levels (168–235 kg ha^{-1}) depressed yields by up to 17%. In Florida, Iley and Ozaki (1967) found maximum yields at 336 kg ha^{-1} N, with reductions of 21% and 33% at higher rates of application. For glasshouse crops, the recommended pre-planting fertiliser dressings range from 105 kg ha^{-1} N for low NO$_3$-N soils (0-25 mg l^{-1} NO$_3$-N) to zero in soils with 51 mg l^{-1} NO$_3$-N or above. The balance of the requirement is supplied by inclusion of 200 mg l^{-1} N

in the irrigation water or by application of 70 kg ha^{-1} N as top dressing at fortnightly intervals (MAFF, 1982).

The form of N supplied may influence the yield, though the results are not always consistent. For instance, Spaldon and Ivanic (1968) obtained higher yields with ammonium sulphate than with ammonium nitrate. In contrast, comparison of ammonium nitrate and ammonium sulphate at three rates of application (56, 140 and 224 kg ha^{-1} N) showed comparable yields at the first two levels but a marked depression by ammonium sulphate at the highest level (Locascio *et al.*, 1981). The highest yields were obtained with a slow release fertiliser, isobutylidene diurea (IBDU). In solution culture, the highest yield was obtained when 10% of the N was supplied in the NH$_4$-N form; increasing the proportion of NH$_4$-N to 40% progressively depressed the yield and increased the incidence of blossom-end rot (Roorda van Eysinga and van der Meijs, 1981).

Nitrogen content of the plants. Miller (1961) reported contents in the vegetative tissue ranging from 1.1% N at low N (42 mg l^{-1} NO$_3$-N in sand culture) to 4.9% at high N (630 mg l^{-1} NO$_3$-N supplied). Deficiency symptoms occurred in plants having a content of 1.26% N or less. Maynard *et al.* (1962) grew pepper plants in sand culture at 100, 200 and 400 mg l^{-1} NO$_3$-N. The leaf N content increased with solution concentration from 2.7% to 3.3% N in one experiment and from 3.9% to 4.8% N in another. The fruit N content ranged from 1.8-2.3% N when grown at 100 mg l^{-1} N in solution to 2.3–2.7% N at 400 mg l^{-1} N. Thomas and Heilman (1964) deduced a critical level of 4.0% N in the leaves and advocated that the N content at the time of initiation of flowering should approach 5%. The N content of plants receiving fairly low applications (0–101 kg ha^{-1} N) decreased markedly during the first part of the growing period, and symptoms of deficiency appeared. The N content of leaves in these treatments increased subsequently and the new foliage thereafter was free from deficiency symptoms even when N had been omitted. Increasing levels of P application decreased the N content of the leaves somewhat and increased the symptoms of N deficiency. Yields were highly and positively correlated with the N content of the leaves. Knavel (1977) similarly reported that yields generally increased with the leaf N levels. Values of 5.4–5.6% N in the leaves were found 4 weeks after transplanting, rising to 6.0–6.4% N at 8 weeks and then declining to 4.2–4.7% N at 12 weeks. A similar pattern of changes during growth, though with lower values throughout, was found by Miller *et al.* (1979). Thus the leaf N values (including petioles) increased from 3.1% when transplanted to a maximum of 5.0% N six weeks later. The values then declined progressively throughout the rest of the season, reaching 2.9% N in the 16th week of growth. A decline in leaf N content of spring-sown peppers, from 4.4% N during May to 3.7% N in June and July, is also evident in the data reported by Locascio *et al.* (1981), these values being the means for 42 combinations of treatments. The yields increased with the N level in the plants. Thus at the final date of sampling, the leaf N values at application rates of 56, 140 and 224 kg ha^{-1} were 3.3%, 3.6% and 4.2% respectively; the corresponding yields were 11.6, 20.3 and 24.2 t ha^{-1}. Sonneveld (1979), studying responses of peppers to salinity in a calcareous loam (5% organic matter), reported 4.6% N in the leaves of the standard plants, 4.4–4.8% N with increased salinity and 5.1% with added sodium nitrate: the nitrogen content of the fruits ranged from 2.5% to 3.1% N. Albasel and Bar-Yosef (1984) grew sweet paprika (*Capsicum annuum* cv. 'Shani') in an inert substrate with various combinations of N concentration and water applications. Plants giving high yields, sampled at harvest time, contained 4.2–4.8% N in the leaves, 1.8–1.9% N in the fruits, 2.0% N in the stems and 1.0–1.3% N in the roots. The total uptake N was 4.5–5.2 g N per plant.

The NO$_3$-N content of pepper plants has received considerable attention, not only as a guide to nutrient status but also with regard to the problem of high levels in food. Stroehlein and Oebker (1979) found decreasing levels of NO$_3$-N both in the leaves and the stems of commercially grown chili pepper at successive dates of sampling. The mean values, each based on 8 combinations of fertiliser treatment, were 0.27%, 0.12% and 0.09% NO$_3$-N in the dried leaves (with petioles). It was concluded that yields might be reduced if the level in the leaves fell below 0.10% NO$_3$-N prior to harvest. Far higher values of up to 1.97% NO$_3$-N were found in stems and petioles (combined samples). Values below 0.4–0.6% NO$_3$-N in the stem tissue at the time of heavy fruit set were insufficient for high yields. Batal and Smittle (1981) found increasing levels of NO$_3$-N in the leaves as the plants matured in the field. The values found (means of four treatments) were 241, 368 and 1023 µg g^{-1} NO$_3$-N in the fresh tissue at 7, 9 and 13 weeks after transplanting.

Detailed information on the NO$_3$-N and total N content of the fruits and vegetative parts of two varieties of sweet pepper, grown with different N sources and concentrations, have been tabulated by Gabal (1983). For example, pepper plants grown at a relatively high N level (30 mg N per 100 g soil, supplied as ammonium nitrate) contained 1.27% NO$_3$-N in the vegetative tops, accounting for 28% of the total N present. The corresponding level in the fruits was only 0.026% NO$_3$-N, equivalent to 0.9% of the total N. Due to nitrification, the levels of NO$_3$-N in the vegetative tissue were also quite high (0.70% NO$_3$-N; 14%

of the total N) even where the N was supplied entirely in the ammonium form (30 mg N per 100 g soil). High day temperatures (25°C compared with 18°C) increased NO$_3$-N concentrations in leaves, stems and fruits. A white-fruited cultivar had appreciably more NO$_3$-N in the fruit than did a red-fruited cultivar. Nevertheless, compared with the vegetative parts of the plant, all NO$_3$-N data for ripe fruit were low.

Phosphorus

Symptoms and growth responses. The leaves of plants deficient in P were narrow, glossy and grey-green in colour, and the overall growth was weak (Miller, 1961). Fruit size and shape were both affected. Thus plants grown in sand culture with 6.2 mg l^{-1} P in solution were reduced in diameter rather than in length, giving a 'pimiento-like' shape (illustrated). More acute deficiency resulted in fruits reduced in both diameter and length, many of which were misshapen (Miller, 1961). Phosphorus deficiency restricted growth, caused yellowing of the leaves, greatly reduced the size and number of fruits formed and delayed harvesting by 10 days (Vereecke, 1975).

Applications of P increased the yield and the proportion of high quality fruit (Ozaki and Iley, 1968), this response being particularly marked on highly organic soils (Bottini, 1968). With a moderate level of application (35 kg ha^{-1} P) the number rather than the size of the fruit increased. There was no further yield response to a higher rate (70 kg ha^{-1} P), but the fruits matured earlier (Pimpini, 1967). Vereecke (1975) grew peppers in peat-filled containers at four levels of P (0, 0.6, 1.8 and 5.4 g l^{-1} P). Maximum yield (by weight) was found at 0.6 g l^{-1} P, with decreases of 5% and 17% at 1.8 and 5.4 g l^{-1} P. The optimal level was considered to be far lower than for other horticultural crops. Fruits grown at the highest level ripened four days earlier.

Phosphorus content of the plants. Miller (1961) noted deficiency symptoms when the vegetative tissue contained 0.09% or less. Plants grown at higher levels of P (31 and 155 mg l^{-1} P in sand culture) contained 0.30% and 0.42% P respectively in the vegetative tissue. Thomas and Heilman (1964) were unable to define a critical P level in the leaves, but concluded that a concentration of 0.6% was adequate. Thus, no increase in yield resulted when the content of the leaves was raised from 0.6–0.7% P to 0.8–1.2% P in response to P application. The P content of the fruits decreased throughout their development, the values ranging from 0.9% in the small fruits (<1.5 cm long) to 0.30% at the mature green stage.

An increase in leaf content from 0.28% to 0.48% P due to P applied to the soil during the previous year was accompanied by increased yields (Ozaki and Iley, 1967). Young pepper plants at the transplanting stage showed levels of 0.21%, 0.31% and 0.58% P after growing with fertiliser applications of approximately 2.5, 25 and 250 kg ha^{-1} P (Ozaki and Iley, 1968). The leaves of the mature plants at the time of first harvest generally showed lower values and far less effect of fertiliser treatment (0.26–0.35% P). Knavel (1977) found significant increases in the P content of pepper transplants with increasing levels of N application in the propagation mixture (peat/vermiculite). Thus the content rose from 0.49% P at 180 g m^{-3} N to 0.56% P at 360 g m^{-3} N under greenhouse conditions and from 0.68% to 0.82% P for similar N treatments after ten weeks in a growth chamber. Strohlein and Oebker (1979) showed a progressive decline in the content of the leaves of field-grown chili peppers, with values of 0.36, 0.27 and 0.19% P at successive sampling dates. Miller *et al.* (1979) followed the changes in P concentration and quantity in the stems, leaves with petioles and fruits throughout the season. The leaf samples showed a progressive decline from 0.43% to 0.27% P during the first ten weeks from planting, after which the concentration rose again to a final value of 0.47% P at 16 weeks. This increase towards the end of the season may result from a diminishing fruit load; up to 58% of the total P in the plants at mid-season was in the fruits. The P in the fruits declined from 0.44% to 0.32% P as the season progressed.

Potassium

Symptoms and responses. The symptoms of K deficiency have been described as 'bronzing' of the leaves. This condition, which appeared shortly after the first harvest of fruit, was accentuated by high levels of N and reduced by applications of K (Osaki and Hamilton, 1955). Symptoms of the deficiency have also been found before the plants reached the flowering stage, accompanied by a marked reduction in growth (Campbell and Swingle, 1966). A severe deficiency resulted in stunted growth, and small necrotic lesions developed along the veins of the bronzed leaves which died prematurely (Miller, 1961).

The highest yields of fruit were generally obtained with intermediate levels of K (Spaldon and Ivanic, 1968). Yields have been depressed by low levels of available K (Ozaki and Hamilton, 1955; Osaki and Ray, 1958; Miller, 1961; Campbell and Swingle, 1966; Iley and Ozaki, 1967) and by heavy application of K (Bottini, 1968; Iley and Ozaki, 1967), though soils with a high organic matter content require heavier dressings than mineral soils (Bottini, 1968). The responses included an increase in the number of fruit and the weight per fruit, in wall thickness (Pimpini, 1967) and in the proportion of high quality fruit (Iley and Ozaki, 1967).

The ratio of K to N taken up, averaged over five varieties of red peppers, was 1.28:1 (Spaldon and Ivanic, 1968). Uptake of K by the plants was greater with ammonium sulphate than it was with ammonium nitrate as the N source.

Potassium content of the plants. Miller (1961) grew peppers in sand culture at concentrations of 0, 47, 235 and 1173 mg l^{-1} K; the corresponding contents of the vegetative tissue (stem and leaves) were 1.0, 1.2, 3.3 and 4.6% K. Omitting the 'nil' treatment, the corresponding values for the fruits were 1.8, 2.9 and 3.4% K. Iley and Ozaki (1967) found relatively low contents in the leaves when sampled at the end of cropping and leaves from plants giving the highest yield contained 2.2% K, rising to 3.2% when grown with a very high level of exchangeable K. Low K application (47 kg ha^{-1} K) resulted in a leaf content of only 0.31% K, accompanied by a reduction in yield of over 40%. Far higher values were reported by Knavel (1977) for pepper transplants grown to first anthesis in growth chambers, ranging from 6.3% to 6.5% K at a day temperature of 24°C and from 6.9% to 7.1% at 27°C. The content of the leaves was found to vary appreciably with time (Miller *et al.*, 1979); values increased from 1.9% K at transplanting to 4.7–5.0% K after 4–8 weeks and then declined to 3.1–3.4% K after 12 weeks growth. Sonneveld (1979) reported 5.7% K in the leaves of sweet peppers grown with a somewhat saline water supply, rising to 7.6% K where excess K was supplied. The corresponding figures for the fruits were 3.9% and 5.2% K respectively. Excess Mg lowered the content of K in the leaves to 4.2%.

Calcium

Symptoms of deficiency. Growth was stunted by Ca deficiency, the leaves became very dark green and the fruit were smaller and darker green than usual (Miller, 1961). The leaves of plants grown for three weeks in sand culture without added Ca were small and yellow, with upturned margins (Hamilton and Ogle, 1962). Many of the fruits on deficient plants showed a disorder similar to blossom-end rot in tomatoes (Miller, 1961). This disorder was usually found near the stylar scar of the fruit (mature or immature) as a grey or whitish sunken area which, though soft at first, hardened as it dried out (Hamilton and Ogle, 1962). In immature fruit the condition spread rapidly and often affected half of the fruit; such fruit usually dropped off. Mature fruit stayed attached to the plant and the green part ripened normally. Factors affecting the incidence of blossom-end rot are discussed later in this section.

Growth responses and fruit quality. The yield of fruit was reduced by a low level of Ca in sand culture (Campbell and Swingle, 1966) and increased by liming a soil which contained only a low level of acid-soluble Ca (Ozaki and Hortenstine, 1962). Heavy liming increased the height of plants growing in a sandy soil, but reduced the yield of fruit (Singh and Nettles, 1962).

Geraldson (1957) found increased incidence of blossom-end rot at low Ca (50 mg l^{-1} Ca) compared with adequate Ca (150 mg l^{-1}) in solution culture. Raising the total salt concentration from 1000 to 3000 mg l^{-1} also increased the amount of blossom-end rot. Even a high level (450 mg l^{-1} Ca) was unable to eliminate the disorder entirely at the higher salinity level. Other workers have confirmed this adverse effect of salinity on the incidence of blossom-end rot, generally attributed to a reduced uptake of Ca and lower content of this element in the fruit walls. Thus, Sonneveld and Voogt (1981) grew peppers in a greenhouse soil irrigated with nutrient solutions having electrical conductivities of 0.45, 0.90, 1.35 and 1.80 mS cm^{-1}; the amounts of blossom-end rot recorded were 1.3, 1.4, 5.6 and 9.3% respectively. Miller (1961) noted that at least half of the fruits produced at low Ca levels in sand culture developed blossom-end rot.

Hamilton and Ogle (1962) grew peppers in sand culture at concentrations of 20, 80 and 160 mg Ca. Some 25% of the fruits produced at low Ca (20 mg l^{-1}) were affected by blossom-end rot, whilst those grown with high Ca (160 mg l^{-1}) were free from the disorder. Application of calcium sulphate to a field soil decreased the incidence of blossom-end rot (Singh and Nettles, 1962); at 560 and 1120 kg ha^{-1} it reduced the numbers of affected fruits by 27% and 35% respectively. Liming an acid soil (pH 4.9) also proved highly beneficial (Ozaki and Hortenstine, 1962). Singh and Nettles (1962) found an adverse effect of fertiliser N on the incidence of blossom-end rot. Raising the nitrogen application from 56 and 168 kg ha^{-1} N by side-dressing with sodium nitrate increased the proportion of affected fruits by 52%. Ammonium nitrogen has a particularly adverse effect on this aspect of fruit quality, as demonstrated by Roorda van Eysinga and van der Meijs (1981). Plants grown in NFT with different proportions of the total inorganic N supplied in the form of NH$_4$–N showed a progressive increase from 3.4% blossom-end rot with 100% NO$_3$–N to 11.2% with 40% NH$_4$–N.

Calcium content of the plants. In several instances, workers have reported Ca levels in both leaves and fruits, relating one or both of these to the incidence of blossom-end rot (BER). Whilst this complicates the overall presentation and the relationships are far from consistent, it nevertheless seems preferable in this section not to separate discussion into leaves and into fruits.

Geraldson (1957) found 1.22% Ca in the leaves of plants grown with adequate Ca (150 mg l^{-1} Ca in solution) compared with only 0.62% Ca at low Ca (50 mg l^{-1}). The corresponding data for the fruits were 0.17% Ca and 0.13% Ca, the latter treatment causing some BER. When B was also omitted from the solutions, the Ca content of the fruit was somewhat higher, and some BER occurred even at 0.20% Ca. Severe BER developed when plants were grown at 150 mg l^{-1} Ca with a high overall salt concentration (3000 mg l^{-1}); the fruits contained 0.16% Ca. Miller (1961) found that fruits containing 0.16% Ca were free from BER whilst those containing 0.08–0.11% Ca were susceptible; the corresponding values for the leaves were 1.69% Ca in the 'standard' plants and

0.32–0.81% Ca in plants showing some BER. Pimiento peppers grown at low Ca in solution (20 mg l^{-1} Ca) produced 25.5% fruits affected by BER; analyses showed 0.18% Ca in the fruits and 0.50% Ca in the leaves (Hamilton and Ogle, 1962). No BER developed in the high Ca treatment (160 mg l^{-1} Ca), corresponding to 0.24% Ca in the fruits and 1.21% Ca in the leaves. Increasing the concentration of Mg in solution over the range 12–97 mg l^{-1} Mg decreased the Ca content of the fruit significantly from 0.23% to 0.19%, though the corresponding increase in BER from 7.7% to 10.9% was not significant. Knavel (1977) grew pepper transplants to first anthesis in growth chambers at day temperatures of 24°C (10 weeks) and 27°C (8 weeks). Leaf analysis showed 1.04–1.29% Ca at the lower temperature compared with 0.77–0.89% Ca at 27°C. Miller *et al.* (1979) recorded changes in the content of Ca and other nutrients in pepper plants throughout cropping, samples of leaves, stems and fruits being collected at fortnightly intervals. Apart from a high value of 2.4% Ca in the leaves (including petioles) at the time of transplanting, the content increased from 1.6% after 4 weeks to 2.2% at 10–12 weeks and then declined to 1.5% at the end of cropping. The content of the fruits decreased with time over the range 0.16% to 0.10% Ca; no reference was made to the incidence of BER. Sonneveld (1979) reported 3.4% Ca in the leaves of sweet peppers, decreasing to 3.0% when the salinity of the irrigation water was increased with sodium nitrate, 2.4% with sodium bicarbonate, 2.2% with magnesium chloride and 1.7% with potassium chloride. Addition of calcium chloride raised the Ca content of the leaves to 4.7% Ca. Fruit analyses showed quite low levels throughout, with 0.06% Ca in the 'standard' treatment, rising to 0.10% Ca with added calcium chloride. All other salinity treatments depressed the content of the fruit to 0.01–0.04% Ca, except for sodium nitrate which increased it to 0.09% Ca. Whilst antagonism between Na and Ca ions might have been expected, the high NO$_3$–N level may have favoured Ca transport. No significant relationship was found by Sonneveld between the Ca content of the fruit and the occurrence of BER. Thus only 1% BER was reported in the 'standard' plants despite there being only 0.06% Ca in the fruits, whereas 16% BER occurred in the high Ca (CaCl$_2$) treatment with 0.10% Ca in the fruits. Nevertheless, the highest incidence of the disorder (24% BER) coincided with the lowest Ca content of all (0.01% Ca) in the fruits.

From the results reported it would seem difficult to establish a 'critical level' for Ca in the fruit, below which BER would necessarily occur. In most experiments relating to BER of peppers, apart from those of Sonneveld (1979), the highest incidence of the disorder tends to coincide with the lowest Ca levels in the fruit. The analytical values vary greatly from trial to trial, however, with BER sometimes reported at quite high Ca levels e.g. 0.20% Ca (Geraldson, 1957). It may be that analyses of whole fruits, as normally reported, do not adequately show the Ca status of the distal region of the fruits where the damage occurs. Alternatively, the Ca status of the fruit, influencing the stability of the cell membranes, may be only one part of a more complex interaction between environmental conditions e.g. water stress and the somewhat delicate fruit tissue. The Ca level below which BER develops may thus rise with the degree of water stress e.g. drought, salinity and air humidity encountered. For further discussion see the corresponding section on blossom-end rot of tomatoes.

Magnesium

Symptoms and growth responses. Out of sixteen crops tested by Carolus (1935), peppers were listed as the easiest crop in which to detect Mg deficiency. The symptoms noted were interveinal yellowing, starting at the tip of the leaves, together with brittleness and an upward leaf curl. Carolus and Brown (1935) stressed the marked contrast between the early yellowing of the interveinal areas and the dark green veins. Growth was restricted, and in cases of severe deficiency the lower leaves dropped off and the fruits were sparse and small. Miller (1961) also described interveinal chlorosis, with the veins and a narrow adjacent portion of the leaf remaining green. Necrotic patches later developed in the chlorotic areas. Both the size and number of fruits were reduced by Mg deficiency, but the shape of the fruit remained normal.

Major yield increases due to Mg application were reported by Carolus and Brown (1935). Thus addition of Mg compounds to a deficient field soil increased the number of fruits produced by 102% and the total yield by 116%. Similarly, Carolus (1935) quoted a yield increase of 51% where Mg was applied. Increased incidence of BER of Pimiento peppers was observed in sand culture as the concentration in solution was raised from 12 to 97 mg l^{-1} Mg, accompanied by a decrease in the Ca content of the fruit (Hamilton and Ogle, 1962).

Magnesium content of the plants. Carolus (1935) reported 0.11% Mg in the leaves of peppers grown in a deficient soil, compared with 0.49% Mg where this nutrient was supplied. The corresponding values for the fruits were 0.13% Mg and 0.19% Mg respectively, i.e. deficiency affected the composition of the leaves more than that of the fruits. Miller (1961) grew bell peppers in sand culture at four concentrations (0, 10, 49 and 243 mg l^{-1}), the third of these levels being regarded as standard and the fourth excessive; the leaf contents were 0.15, 0.20, 0.60 and 1.37% Mg respectively. Omitting the 'nil' treatment, the corresponding values for the fruits were 0.20, 0.22 and 0.27% Mg; again, the fruits showed less variation than did the leaves. Healthy leaves were found to contain 0.65-0.72% Mg (Ozaki and Iley, 1968) and no

positive yield responses to applied Mg were found in these trials. Deficient (chlorotic) leaves of Pimiento peppers contained 0.25% Mg compared with 0.58% Mg in healthy leaves (Dempsey and Brantley, 1969). Knavel (1977) recorded higher levels in pepper transplants when grown to first anthesis at a day temperature of 24°C (1.4–1.7% Mg) than at 27°C (0.9–1.1% Mg). Miller et al. (1979) showed a progressive increase in leaf content from 0.6% Mg at the time of transplanting to 1.6% Mg after 12 weeks. The level then declined, reaching 0.9% Mg at 16 weeks. Magnesium concentrations in the stems followed a similar pattern, rising from 0.6% to 1.0% Mg and then declining to 0.5% Mg at the end of the trial. The fruits, however, showed a small but progressive decrease with time, from 0.24% Mg to 0.18% overall. Some 50–80% of the total Mg in the plant was located in the leaves (plus petioles), 18–38% in the stems and up to 16% in the fruits. Additions of Na, K or Ca salts as sources of salinity in the irrigation water depressed the Mg content of the leaves from 1.48% to 1.0–1.3% Mg, whereas the addition of a Mg salt increased the Mg content of the leaves to 3.1% Mg (Sonneveld, 1979). Locascio et al. (1981) found 0.47–0.58% Mg in whole pepper plants at the time of first setting.

Iron

A chlorosis of the leaves of peppers grown in a peat/vermiculite mixture (Cornell Peat-lite Mix A) was shown to be due to Fe deficiency induced by the high Mn:Fe ratio of the vermiculite (Read and Sheldrake, 1966). Spraying with chelated Fe corrected the condition, with Fe-DTPA acting more rapidly than Fe-EDDHA. Hewitt and Watson (1980) grew peppers without added Fe in re-circulating solution culture. The symptoms included yellowing of the newly expanded leaves, leaf scorch and arrested growth. The roots were particularly stunted and blackened at the tips. A yellow-fruited inbred pepper cultivar ('Zehavi') became chlorotic when grown in a previously uncultivated calcareous soil (Shifriss and Eidelman, 1983). Diagnostic tests indicated Fe deficiency, and the condition was corrected by application of chelated iron (Fe-EDTA) to the soil. Inheritance studies indicated that susceptibility to Fe stress was under recessive control. Navrot and Levin (1976) reported 82–98 µg g^{-1} Fe in the fruits, the mean values showing no response to treatment with Fe compounds.

Manganese

Manganese deficiency in field-grown peppers in a calcareous Glade soil in Florida was described by Skinner and Ruprecht (1930). The leaves developed a severe interveinal chlorosis (illustrated), growth was much restricted and yields were poor. Applications of manganese sulphate at 73 and 140 kg ha^{-1} increased the yields by 70% and 81% respectively. Manganese deficiency caused yellow spots to develop in the older leaves of pepper plants in re-circulating solution culture; the affected areas later became necrotic (Hewitt and Watson, 1980).

Copper

Copper deficiency causes rolling of the young leaves, which later wilt and dry out (Rahimi, 1970; illustrated). Green pepper sown without added Cu made little growth (Hewitt and Watson, 1980). Plants transferred from complete nutrient solution to conditions of Cu deficiency showed wilting of the leaf margins, and the young, dark green leaves developed a faint mottling (Vol.1, Plate 41). Navrot and Levin (1976) recorded 23 μg g^{-1} Cu in the fruits from field-grown plants without added Cu, and 26-31 μg g^{-1} where this element was supplied.

Zinc

Pepper plants grown at a low level of Zn in re-circulating solution culture (Hewitt and Watson, 1980) showed interveinal bronzing of the leaves and dehiscence of the pedicels (Vol. 1, Plate 47). Navrot and Levin (1976) recorded 44-45 μg g^{-1} Zn in the fruits of field-grown plants without added Zn and 43-52 μg g^{-1} where this element was supplied.

Boron

Eaton (1945) observed no symptoms of deficiency in peppers grown at a 'trace' level of B (0.03-0.04 mg l^{-1}) in out-door sand culture. At 5 mg l^{-1} B the margins of the older leaves were restricted and down-curled. The toxicity symptoms were more acute at 10 and 15 mg l^{-1} B, and at 25 mg l^{-1} the leaves were much reduced in size, with severe marginal scorch (illustrated, Eaton, 1945). The total dry weights of the plants were reduced by 8–9% at 1 and 5 mg l^{-1} B, and by 28%, 51% and 71% respectively at 10, 15 and 25 mg l^{-1} B; the corresponding decreases in the yields of fruit were 24%, 51% and 85% at 10, 15 and 25 mg l^{-1} B. The B contents of the leaves (μg g^{-1} B), with solution concentrations in mg l^{-1} B shown in parentheses, were 34 (0.03–0.04), 118 (1), 328 (5), 700 (10), 729 (15) and 882 (25). Fruits grown with 1–5 mg l^{-1} B in solution contained 16–21 μg g^{-1} B, rising to 46 μg g^{-1} at the highest concentration.

Hewitt and Watson (1980) recorded that peppers deprived of B were stunted; the youngest leaves bent sideways and the pedicels dehisced.

Molybdenum

The leaves of peppers grown in re-circulating nutrient solution without added Mo were described as irregularly notched and mottled along the margins (Hewitt and Watson, 1980).

Micronutrient combinations

Navrot and Levin (1976) applied foliar sprays containing B, Cu + Zn, B + Cu + Zn and B + Cu + Zn + Fe as inorganic salts, and also a proprietary micronutrient mixture containing B together with Cu, Zn, Mn and Fe in chelated (EDTA) form. Applications of B + Cu + Zn to the soil (a drained peat) were also tested, with and without added Fe (FeSO$_4$). Only the proprietary formulation containing nutrients in chelated form gave a significant increase in yield (21%). No specific symptoms of deficiency were seen in any of the plots. It was noted, however, that whereas plants sprayed with the chelated mixture maintained their deep green colour, plants in the other treatments showed some yellowing at the onset of flowering.

References:

Albasel, N. and Bar-Yosef, B. (1984). Effect of N solution concentration, water rate and transient starvation in sweet paprika dry matter production and yield. *J Pl Nutr* **7**, 1005–1018.

Batal, K. M. and Smittle, D. A. (1981). Response of bell pepper to irrigation, nitrogen, and plant population. *J Am Soc hort Sci* **106**, 259–262.

Bottini, E. (1968). Mineral fertilisation of peppers. *Pot Symp 1966*, 293-301.

Campbell, G. M. and Swingle, H. D. (1966). N-K teamwork peps up sweet pepper yields. *Better crops* **49**, No. 6, 30–32.

Carolus, R. L. (1935). Effects of magnesium deficiency in the soil on the yield, appearance and composition of vegetable crops. *Proc Am Soc hort Sci* **32**, 610–614.

Carolus, R. L. and Brown, B. E. (1935). Magnesium deficiency. I. The value of magnesium compounds in vegetable production in Virginia. *Bull Va Truck Exp Stn* **89**, 1250–1288.

Cochran, H. L. (1936). Some factors influencing growth and fruit-setting in the pepper (*Capsicum frutescens* L.). *Cornell agric Expt Stn Mem* **190**, 29pp.

Dempsey, A. H. and Brantley B. B. (1969). Magnesium deficiency symptoms and related leaf tissue composition of pimiento pepper. *Res Rep Ga Coll agric Expt Stn* **52**, 6pp.

Eaton, F. M. (1945). Deficiency, toxicity and accumulation of boron in plants. *J agric Res* **69**, 237–277.

Gabal, M. R. (1983). Effect of N-doses, N-form and day temperature on nitrate accumulation in sweet peppers. *Acta agron Hung* **32**, 377–387.

Geraldson, C. M. (1957). Factors affecting calcium nutrition of celery, tomato and pepper. *Soil Sci Soc Am Proc* **21**, 621–624.

Hamilton, L. C. and Ogles, W. L. (1962). The influence of nutrition on blossom-end rot of pimiento peppers. *Proc Am Soc hort Sci* **80**, 457–461.

Hand, D. W. (1980). Fuel savings for early heated crop. *Grower* **93** (BGLA Supplement), 21 February, 75, 77.

Hewitt, E. J. and Watson, E. F. (1980). The production of micronutrient element deficiencies in plants grown in recirculated nutrient film (NFT) systems. *Acta Hort* **98**, 179–189.

Iley, J. R. and Ozaki, H. Y. (1967). Nitrogen-potash ratio study with plastic mulched pepper. *Proc Fla St hort Soc* **79**, 211–216.

Knavel, D. E. (1977). The influences of nitrogen on pepper transplant growth and yielding potential of plants grown with different levels of soil nitrogen. *J Am Soc hort Sci* **102**, 533–535.

Locascio, S. J., Fiskell, J. G. A. and Martin, F. G. (1981). Responses of bell pepper to nitrogen sources. *J Am Soc hort Sci* **106**, 628–632.

Maynard, D. N., Lachman, W. H., Check, R. M. and Vernell, H. F. (1962). The influence of nitrogen levels on flowering and fruit set of peppers. *Proc Am Soc hort Sci* **81**, 385–389.

Miller, C. H. (1961). Some effects of different levels of five nutrient elements on bell peppers. *Proc Am Soc hort Sci* **77**, 440–448.

Miller, C. H., McCollum, R. E. and Claimon, S. (1979). Relationships between growth of bell peppers *(Capsicum annuum* L.) and nutrient accumulation during ontogeny in field environments. *J Am Soc hort Sci* **104**, 852–857.

Ministry of Agriculture, Fisheries and Food. (1981) Sweet Peppers. Booklet **2190** (formerly STL **818**), 17pp.

Ministry of Agriculture, Fisheries and Food. (1982). Lime and fertilisers recommendations No. 4. Glasshouse crops and nursery stock 1983/84. Booklet **2194**, 48pp.

Navrot, J. and Levin, I. (1976). Effect of micronutrients on pepper *(Capsicum annuum)* grown in peat soil under greenhouse and field conditions. *Expl Agric* **12**, 129–133.

Ozaki, C. T. and Hamilton, M. G. (1955). Bronzing and yield of peppers as influenced by varying levels of nitrogen, phosphorus and potassium fertilisation. *Proc Soil Sci Soc Fla* **14**, 185–189.

Ozaki, H. Y. and Hortenstine, C. C. (1962). Effect of lime on yield of peppers and on soil calcium. *Proc Soil Sci Soc Fla* **21**, 50–55.

Ozaki, H. Y. and Iley, J. R. (1967). Nutritional requirements of vegetable crops grown on sandy soils in Florida. *Fla agric Expt Stn Ann Rep 1966*, 304.

Ozaki, H. Y. and Iley, J. R. (1968). Phosphorus and magnesium fertiliser studies with pepper. *Proc Am Soc hort Sci* **93**, 462–469.

Ozaki, H. Y. and Ray, H. E. (1958). Fertiliser studies of vegetable crops grown on the sandy soils of the lower East coast. *Fla agric Expt Stn Ann Rep 1957*, 291–292.

Pimpini, F. (1967). Experiments with the mineral fertilisation of sweet peppers. *Progr agric, Bologna* **13**, 915–932.

Rahimi, A. (1970). Kupfermangel bei höheren Pflanzen. *Landwirt Forsch Sonderh* **25**, 42–47.

Read, P. E. and Sheldrake, Jr., R. (1966). Correction of chlorosis in plants grown in Cornell peat-lite mixes. *Proc Am Soc hort Sci* **88**, 576–581.

Roorda van Eysinga, J. P. N. L. and van der Meijs, M. Q. (1981). Disorders in red sweet pepper fruits. *Ann Rep Glasshouse Crops Res Exp Stn, Naaldwijk, 1979*, 24–25.

Shifriss, C. and Eidelman, E. (1983). Iron deficiency chlorosis in peppers. *J Pl Nutr* **6**, 699–704.

Singh, K. and Nettles, V. F. (1962). Effect of defloration, defruiting, nitrogen and calcium on the growth and fruiting of bell peppers *(Capsicum annuum* L.) *Proc Fla St hort Soc* **74**, 204–209.

Skinner, J. J. and Ruprecht, R. W. (1930). Fertiliser experiments with truck crops. III. Truck crops with manganese on calcareous Glade soil. *Fla agric Exp Stn Bull* **218**, 37–65.

Sonneveld, C. (1979). Effects of salinity on the growth and mineral composition of sweet pepper and eggplant grown under glass. *Acta Hort* **89**, 71–78.

Sonneveld, C. and Voogt, S. J. (1981). Nitrogen, potash and magnesium nutrition of some vegetable fruit crops under glass. *Neth J agric Sci* **29**, 129–139.

Spaldon, E. and Ivanic, J. (1968). The role of potash in the nutrition of the red pepper *(Capsicum annuum)*. *Pot Rev, Subj 16, Suite* **40**, 1–14.

Stroehlein, J. L. and Oebker, N. F. (1979). Effects of nitrogen and phosphorus on yields and tissue analyses of chili peppers. *Commun Soil Sci Pl Anal* **10**, 551–563.

Thomas, J. R. and Heilman, M. D. (1964). Nitrogen and phosphorus content of leaf tissue in relation to sweet pepper yields. *Proc Am Soc hort Sci* **85**, 419–425.

Vereecke, M. (1975). Phosphorus fertiliser studies with *Capsicum annuum* L. (sweet pepper). *Acta Hort* **50**, 83–87.

CHAPTER 5

Tomato

The tomato (*Lycopersicon esculentum* Mill) crop is usually well supplied with fertilisers (see p. 7), and many of the commoner nutritional disorders arise indirectly from cultural practices. For example, Mn toxicity may be caused by partial steam-sterilisation of many soils, and Mg deficiency is frequently induced by the heavy applications of K used to ensure that the fruit are of good quality. Other disorders arise because the watering is not controlled closely enough; under-watering effectively increases the salinity of the substrate and reduces the availability of some nutrients, whilst overwatering results in poor aeration, root death and loss of nutrients by leaching.

The nutrient status is also affected by other factors such as the form of N applied, the environment and interactions between nutrients. Thus, blossom-end rot (BER), which occurs when the fruit have a very low Ca content, has been associated with several different conditions, including an acidic (Ca deficient) substrate, high levels of NH_4-N or excessive amounts of K in the substrate, soil salinity or dryness and low temperature. As a further example of the influence of plant environment, some effects of light intensity (position in glasshouse) on nutrient uptake by tomatoes and on the percentages of dry matter and nutrients in the leaves are shown in Table 10. Plants in the outer rows, close to the glass, showed higher total uptakes of nutrients (N, P and K) than those in the inner rows. These outer plants, exposed to higher levels of solar radiation, had a higher percentage of dry matter in their leaves but lower percentages of a nutrients when expressed on a dry matter basis.

	Nutrient uptake (g/plant)			% dry matter in leaves	% nutrient in leaves		
	N	P	K		N	P	K
Outer rows (exposed)	17.0	1.42	28.3	11.42	4.64	0.24	4.46
Inner rows (mutual shading)	13.3	1.38	22.8	9.34	5.25	0.32	5.20

Table 10. Effect of position in the glasshouse on total nutrient uptake by tomato plants, % dry matter in the leaves and % nutrients expressed on the basis of leaf dry matter (based on Winsor, Davies and Messing, 1958).

Nitrogen

Symptoms and growth responses. The N status of the crop is reflected very clearly in the growth and colour of the plants. Nitrogen deficiency results in stunted growth, the leaves being very small, stiff and pale yellow-green. The older leaves become yellow with pink veins and die prematurely. The flower buds turn yellow and fall off. Descriptions of N deficiency have been given by Fisher (1935), Hewitt (1944), Woolley and Broyer (1957) and others. The condition was illustrated in colour for young tomato plants by Wallace (1951), and for mature plants by Smilde and Roorda van Eysinga (1968), Roorda van Eysinga and Smilde (1981).

Responses to N have been reported widely, but these vary according to the nature and N content of the substrate, the amount and form of N applied and other interacting factors such as the mode of culture, the environment and partial sterilisation of the substrate. Favourable responses are usually recorded to increasing amounts of available N. However, both the growth and yield may be depressed by excessive amounts of NO_3-N, by moderate amounts of NH_4-N and by small amounts of NO_2-N or free ammonia.

White (1938) showed that plant height and leaf area increased with the level of N applied; deficiency reduced both the number of flower buds formed and the proportion of buds which opened, delayed flowering and also increased the time required for the fruit to mature and ripen. The yield of fruit generally increases with the rate of N applied (Stromme, 1957; see also Fig. 13), but may be depressed by very high levels (Malcolm, 1959), particularly when the K level is inadequate or when the soil is heavily limed (Winsor *et al.*, 1967). The incidence of ripening disorders is usually highest at intermediate levels of N (Adams *et al.*, 1978a; see Fig. 14), but may increase at high levels when the K status of the substrate is inadequate (Winsor and Long, 1967).

Effects of ammonium-N. Depressions in the rate of vegetative growth due to high levels of NH_4-N have been reported by several workers, including Ulgee (1964) and Torres de Claasen and Wilcox (1974). The growth of young plants may be markedly affected by an interaction between the rate and form of N applied and the environment. Thus Bunt (1969) grew tomato seedlings in a peat/sand compost at four levels of N supplied as a mixture of hoof and horn and ammonium nitrate. The highest rate of N depressed growth during the winter whereas a positive response was found in the summer. Depression of growth during winter is usually due to the persistence of appreciable amounts of NH_4-N, and the best growth is

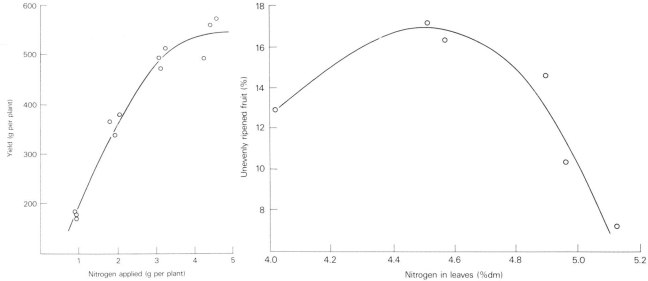

Figure 13 Relation between the amount of nitrogen applied to single-truss tomato plants and the yield of fruit (after Adams *et al.*, 1973).

Figure 14 Relation between the nitrogen content of tomato leaves and the proportion of unevenly ripened fruit (after Adams *et al.*, 1978a).

obtained using NO_3-N. Thus the fresh weights of tomato seedlings grown for 7 weeks in a peat-sand compost varied with N source as follows: 1.18 g with urea, 5.54 g with ammonium sulphate and 8.49 g with calcium nitrate (Bunt, 1976). The depression in growth associated with the mineralisation of an organic nitrogenous fertiliser becomes greater as the pH of the soil increases (see Court *et al.*, 1964 a and b), and is particularly marked in soil which has been partially sterilised by steaming (Bunt, 1976; see also Table 11). The release of ammonia from nitrogenous organic compounds can itself raise the pH of the substrate appreciably during the first week or so and thus aggravate the problem by increasing the proportion of N present as free ammonia.

The first symptom of ammonia toxicity in tomatoes is usually yellowing of the leaves (Puritch and Barker, 1967), but dark pitted lesions develop on the stems when the toxicity becomes severe (Barker *et al.*, 1967). The incidence and severity of the lesions may be greatly reduced by increasing the K status of the plants (Maynard *et al.*, 1968). The use of relatively high levels of K in substrates for commercial tomato production has undoubtedly minimised the effects of what might otherwise have often been a serious problem.

	Initial soil pH	
	5.2	6.3
Unsteamed soil	1.10	0.70
Steamed soil	1.02	0.12

Table 11. Effect of partial steam sterilisation and liming on the dry weight (g) of tomato seedlings grown for eight weeks during winter in soil treated with an organic nitrogenous fertiliser (hoof and horn; A.C. Bunt, private communication).

	% nitrogen supplied as ammonium-nitrogen		
	0	20	40
Fruit affected by blossom-end rot in the first four harvests (% by number)	0	24	46
Calcium content of leaves (% Ca)	1.82	1.48	0.87

Table 12. Effect of ammonium-nitrogen on the proportion of fruit affected by blossom-end rot and on the calcium content of the leaves of tomato plants grown in nutrient film culture (after Massey and Winsor, 1980a).

The use of NH_4-N depresses the uptake of other cations such as K, Ca and Mg (Kirkby and Mengel, 1967) and may also induce BER of the fruit (see p.59). Massey and Winsor (1980a) varied the proportion of the N supplied as NH_4-N to tomato plants in recirculating solution culture. The effect of the treatments on the incidence of BER and the Ca content of the leaves is shown in Table 12.

Nitrogen content of the leaves. The total N content of the young mature leaves (usually the fifth leaf below the head of the plant) is widely used to assess the current status of the crop. Several estimates of the N content of the leaves corresponding with maximum yield have been made, but the values suggested vary considerably, probably due to the diversity of the experimental conditions concerned. For example, van der Kloes *et al.* (1961) and Roorda van Eysinga (1971a) suggested 3.5% and 3.7% N respectively, whereas Ward (1963) has recommended 5.25% N. Trials with tomatoes grown in beds of peat suggested that the highest yields would be obtained when the leaves contained 4.5-5.1% N (Adams *et al.*, 1978b). A similar value (4.9% N) was obtained with single–truss tomatoes (Adams *et al.*, 1973). As shown in Figure 14, values of 4.9-5.1% N in the leaves were accompanied by reduced incidence of ripening disorders, probably due to increased salinity. Published values associated with healthy growth include 4.3-6.0% N (MacLean *et al.*, 1968), 4.8% N (Winsor, 1973), 4.0-5.6% N (Winsor and Massey, 1978) and 4.0-5.5% N (Sheldrake, 1981) compared with less than 1.7% N in deficient plants (Roorda van Eysinga and Smilde, 1981). Starck *et al.* (1973) found a progressive decline in the N content of young fully expanded leaves of tomato plants grown in peat, the values ranging from about 5.0-5.5% N in March to about 2-3% N in June. The effect of plant age was generally greater than that of N application.

Several authors have suggested that the amount of NO_3-N that has accumulated in the leaves gives a better estimate of the current status of the crop than does the total N content; see, for example, Mason and Wilcox (1982). Roorda van Eysinga (1971a) found that 0.73% NO_3-N corresponded with maximum yield, and N deficiency occurred at contents of less than 0.1% NO_3-N. Healthy plants normally contained 0.5-1.5% NO_3-N in their leaves (Smilde and Roorda van Eysinga, 1968). Gomez-Lepe and Ulrich (1974) suggested that the dried petioles from plants receiving an adequate supply of N contained 0.1-4.0% NO_3-N. Using the petiole of the second leaf below the head, these authors proposed a critical value of 0.05% NO_3-N in the dry matter for use under field conditions. The NO_3-N content of the leaves increased with age and was higher in the petioles than in the laminae.

Phosphorus

Symptoms and growth responses. Phosphorus deficient tomato plants were described by Fisher (1935) as very dark blue-green or purplish in colour. The stems were thin and stunted, and the roots became brown and developed few lateral branches. Hewitt (1944) noted that the mature leaves were small, with down-curled leaflets. The oldest leaves, which had purple tints and scorched areas, became yellow with purple veins and died prematurely. A similar description was given and illustrated by Wallace (1951). Purpling on the undersides of the leaves included both the veins and interveinal areas (Woolley and Broyer, 1957).

Much attention has been given to maintenance of the P status of glasshouse soils, and marked responses in both growth e.g. Martin and Wilcox, 1963 and the yield of fruit (Winsor *et al.*, 1967) have been recorded. These responses may be modified, however, by several factors that affect the availability of P to the crop including the pH, moisture content, degree of compaction and temperature of the substrate as well as the volume of substrate per plant.

The availability of P declines as the pH of the substrate increases. Thus, heavy liming reduced the uptake by tomato plants (Fig. 15) and had an adverse effect on growth and yield (Winsor and Long, 1963). Liming the soil

from pH 6.3 to 7.4 depressed the yield by 10–16% (Winsor *et al.*, 1967) and increased the proportion of unevenly ripened fruit, particularly at high levels of N (Winsor and Long, 1967). Firmness of the fruit was improved, however, at the higher pH (Shafshak and Winsor, 1964).

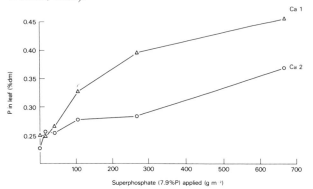

Figure 15 Effects of superphosphate and liming on the phosphorus content of tomato leaves (Ca 1, △; Ca 2, 0; Winsor and Massey, unpublished data).

In peat or peat-based substrates a high proportion of the P applied may remain water-soluble and, unless the watering is controlled carefully, much of this may be lost by leaching. When the crop is grown in bags of peat (8–10 l per plant), P is often therefore included in the liquid feed 4–8 weeks after planting. The amount of water-soluble P in fertilised peat is closely related to pH. For example, over 80% of the 'available' (acid-soluble) P was water-soluble at pH 5, but less than 20% remained water-soluble at pH 7.0. Increasing the pH of a peat substrate from 4.6 to 6.2 reduced the P content of the leaves by 30% (Adams, 1978a).

Phosphorus diffuses very slowly in soil. The uptake of sufficient P to support normal growth is thus dependent on continuous root extension, the rate of which is determined partly by the degree of compaction of the soil. Menary and Kruger (1966) showed that the uptake declined as the bulk density of the soil increased due to compaction. At densities greater than 1.3 g cm^{-3} the plants were stunted, and the purple colouration associated with deficiency was found on the undersides of the leaves and on the petioles and stems.

In solid substrates, the soil solution around the roots is rapidly depleted of P, whereas in nutrient film culture the roots are continuously bathed in a flowing nutrient solution at a maintained concentration. For this reason, relatively low concentrations suffice in nutrient film culture. Thus Massey and Winsor (1980b) found no significant yield responses by tomato over the range 5–200 mg l^{-1} P and even at the lowest of these concentrations the leaves contained at least 0.43% P.

The uptake of P generally increases with the temperature of the roots. For example, a marked increase in the P content of the leaves was found when the temperature of the soil was raised over the range 13.3° to 15.5°C (Martin and Wilcox, 1963). Locascio and Warren (1960) found a 15–fold increase in the total uptake when the soil temperature was increased from 13° to 21°C, with a further increase of 15% at 30°C. Much of the increase in uptake between 13° and 21°C was attributed to increased root growth rather than to increased availability of P. In nutrient film culture, the P content of the leaves increased by 20% when the temperature of the nutrient solution was maintained at 25°C instead of being left to fluctuate in response to changes in air temperature and solar radiation (Maher, 1978).

Phosphorus content of the leaves. Roorda van Eysinga (1971b) concluded that there was no yield response to increases in the P content of tomato leaves above 0.44% P in the dry matter. The relationship between yield and %P in the leaves, shown in Figure 16, indicates the importance of maintaining adequate levels of P for tomatoes; little or no further response would be expected above 0.45% P. Estimates of leaf P regarded as satisfactory or optimal include 0.39% P (Wallace, 1951), 0.80% P (Ward, 1963), 0.43−0.60% P (Smilde and Roorda van Eysinga, 1968), 0.50% (Winsor, 1973), 0.4% or more (Besford, 1979b) and 0.41% (Swiader and Morse, 1982). Analyses of the laminae of young tomato plants (52 days from sowing), grown in solution culture, showed 0.51, 0.74, 0.86 and 1.25% P at root temperatures of 12°, 15°, 18° and 30°C respectively (Cornillon, 1974).

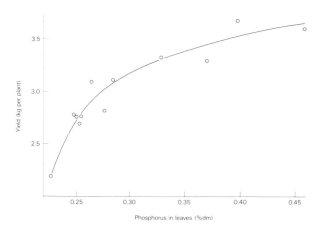

Figure 16 Relation between the phosphorus content of tomato leaves and the yield of fruit (Winsor and Massey, unpublished data).

Concentrations of 0.11−0.16% P (MacGillivray, 1927), or less than 0.17% P (Smilde and Roorda van Eysinga, 1968) and of less than 0.13% P in the leaves (Besford, 1979a) have been associated with deficiency. The growth of young plants was restricted when the leaves contained less than 0.19% P (Hogue *et al.*, 1970). Bradley and Fleming (1960) found highly significant positive correlations (P < 0.01) between yield and leaf P early in the

season for field grown tomatoes. Cannell *et al.* (1960) grew young tomato plants (< 8 weeks old) at four P levels in three soils. Shoot growth (by weight) increased progressively up to the highest rate of supply in each soil, corresponding to leaf P levels of 0.79%, 0.82% and 0.90% P. Swiader and Morse (1982) reported a yield decrease of 73% from field-grown tomatoes at the highest rate of P application in their trials, corresponding to 0.72% P in the leaves. This response, which was shown not to be due to 'P–induced' Zn deficiency, was accompanied by marginal necrosis which started at the tip of the older leaves.

Sulphur

Sulphur deficiency is seldom found in vegetable crops grown within several hundred miles of industrial centres (Purvis and Carolus, 1964). Roorda van Eysinga and Smilde (1981) stated that this deficiency was unknown in commercial glasshouses, most of which receive large amounts of SO_4-S from fertilisers such as potassium sulphate, magnesium sulphate and superphosphate (containing calcium sulphate). The symptoms have, however, been studied in sand and solution culture by several workers, including Nightingale *et al.* (1932), Woolley and Broyer (1957), Smilde and Roorda van Eysinga (1968), Ward (1976) and Roorda van Eysinga and Smilde (1981).

Nightingale *et al.* (1932) noted some general similarity to N deficiency; 'the plants looked as though they had been gradually but not completely deprived of N'. The lower leaves were yellowish green, and the stems hard and woody. Smilde and Roorda van Eysinga (1968) described the older leaves as showing necrosis at the tips and margins, with small purple spots between the veins. The young leaves curled downwards and later developed irregular necrotic patches. Ward (1976) found it difficult to define a 'range of sufficiency', but tissue levels below 0.25% S indicated severe deficiency. Roorda van Eysinga and Smilde (1981) recorded a range of 0.96−3.20% S in healthy tomato leaves, with deficiency below 0.48% S.

Potassium

Potassium is not only essential for normal plant growth, but is undoubtedly the most important nutrient affecting the quality of tomato fruit. It is now well established that the amounts required for production of evenly ripened fruit considerably exceed those required for maximum yield (Roorda van Eysinga, 1966; Winsor and Long, 1967; Adams *et al.*, 1978a).

Symptoms of deficiency. Bewley and White (1926) recorded that the foliage of K deficient tomato plants was at first abnormally dark green, and frequently purplish brown in colour. Marginal chlorosis developed later and spread interveinally; the leaf margins eventually turned brown. Omission of K fertiliser increased the incidence of blotchy ripening, averaged over five seasons, from 2.9% to 19.9% in one set of plots and from 3.2% to 19.4% in another. Johnston and Hoagland (1929) described and illustrated somewhat different symptoms of deficiency. Yellowish spots appeared, at first near the leaf margins but later scattered over the whole surface of the leaf; these spots rapidly turned dark brown, starting with the older leaves. Similar symptoms, starting with yellow-brown marginal spotting, were reported by Fisher (1935). Wall (1940) noted two stages in the development of deficiency symptoms in sand culture. Initially the deficient plants were stunted, hard and chlorotic. Later, however, the plants became green and began to grow, whilst the lower leaves began to die progressively up to the stem. White (1938) confirmed the dark green colour of the early stages of deficiency in soil-grown plants, later followed by marginal chlorosis. Hewitt (1944) recorded that the lower leaves became grey at the margins and later interveinally; the tips and margins became scorched and curled upwards. Eventually the leaves at the top of the plant became severely scorched. Most descriptions and illustrations e.g. Wallace, 1951; Smilde and Roorda van Eysinga, 1968; Bergmann, 1976, suggest that the margins first became yellow rather than grey, and later turn brown. Woolley and Broyer (1957) recorded tip and marginal chlorosis on the older leaves of mildly affected plants. Old or recently matured leaves showed interveinal chlorosis, and small interveinal necrotic spots sometimes developed on young leaflets cf. Johnston and Hoagland, 1929. The fruit ripened unevenly and were insipid, lacking acidity (Hewitt, 1944). Wallace (1951) described the fruit as having 'green and yellow areas merging into the red colour of the surface'. The histology of tomato fruit affected by blotchy ripening, including the severe form with clear 'glassy' blotches accompanied by vascular browning, was examined by Seaton and Gray (1936); see also earlier observations by Bewley and White (1926). The disorder was reviewed and illustrated by Hobson et al. (1977), with recommendations for its control including the maintenance of high levels of soil K.

Growth responses Plant height and leaf area were depressed by K deficiency (White, 1938). Flower formation was unaffected, but the numbers of fruits developing per truss were decreased. The proportion of fruit which set and the average weight per fruit were also reduced by deficiency (Clarke, 1944). Yield responses have been somewhat variable. Several authors e.g. White, 1938; Malcom, 1959, reported an increase in yield with increasing levels of K whereas Stromme (1957) found no response by a soil-grown crop; the initial level of soil K doubtless influences the results of such trials. Interactions with other nutrients are important, however, and the increase in yield with K is greatest when the level of N is not limiting (Winsor et al., 1967). A reduction in yield is frequently found at very high levels of K (Walsh and Clarke, 1945a;

Adams et al., 1978b), with an increasing K:N ratio in the liquid feed (Winsor et al., 1961) and with a combination of high levels of both N and K (Winsor et al., 1962). This reduction in yield has been attributed to the effects of increased salinity in the substrate (Clay and Hudson, 1960). High levels of K may affect yield or fruit quality indirectly by depressing the Mg and Ca status of the plants. A chlorosis of tomato leaves described and illustrated by Walsh and Clarke (1942) was associated with excessively high levels of soil K.

Potassium and fruit quality. Increasing levels of K generally improve fruit shape, decrease fruit size (Winsor and Long, 1968) and reduce the incidence of ripening disorders (Bewley and White, 1926; Winsor and Long, 1967; Winsor, 1970; Adams et al., 1978a; see Fig. 17). Some examples quoted by Winsor (1979) are shown in Table 13. Fruit firmness is improved by high levels of K (Shafshak and Winsor, 1964), and the proportion of hollow fruit, a disorder associated with early growth under poor light conditions, is greatly reduced (Winsor, 1966).

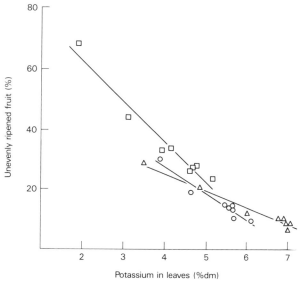

Figure 17 Relation between the potassium content of tomato leaves and the proportion of unevenly ripened fruit (□, Crop 1; 0, Crop 2; △, Crop 3; after Adams et al., 1978a).

	Potassium applied (kg ha^{-1})			
	359	706	1428	LSD (P=0.05)
Weight (g) per fruit	78.9	75.3	71.5	2.9
% Irregularly shaped (amongst the uniformly coloured fruit)[+]	56.3	32.6	28.0	2.9
% Unevenly ripened[+]	40.5	12.4	5.8	3.3
% Unevenly ripened[+] (severe forms only)[+]	24.1	5.3	1.3	2.6

Table 13. Some effects of potassium on the size and shape of tomato fruit and on the incidence of ripening disorders (cv. J352; after Winsor, 1979). [+] Percentage by weight.

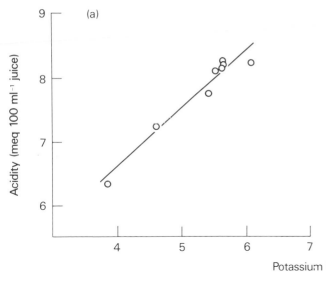
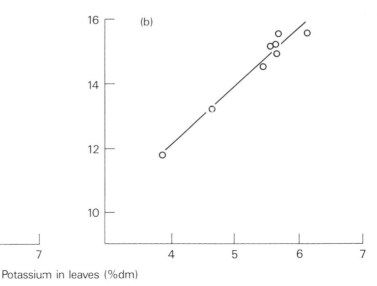

Figure 18 Relation between the potassium content of tomato leaves and
(a) the titratable and (b) total acidities of the fruit juices (after Adams *et al.*, 1978a).

The free (titratable) acidity of the fruit is one of the main taste components of the tomato and is closely related to the K status of the soil (Davies, 1964; Davies and Winsor, 1967) and of the plants (Adams *et al.*, 1978a; see Fig. 18). Potassium is the predominant cation in tomato fruit and the combined acidity, calculated as the difference between total and titratable acidity, is particularly well correlated with K concentration. Based on a total of 288 samples (Davies and Winsor, 1967), the regression of combined acidity (x) on K concentration (y), both expressed as meq 100 ml^{-1} of tomato juice, is given by the equation x = 0.98 y + 1.25 (r = 0.94).

Potassium content of the leaves. Reported values for the K content of tomato leaves associated with deficiency show considerable variation. For example, Wallace (1951) quoted 1% K and Roorda van Eysinga and Smilde (1981) suggested less than 1.17% K. Besford and Maw (1975) found deficiency symptoms in two cultivars when the leaves contained less than 1.2–1.5% K, whilst Bergmann (1976) reported only 0.54% K. These values probably reflect the requirements of the plants when making mainly vegetative growth. Plants bearing a heavy load of fruit have a much greater demand, however, as more than half of the K absorbed by the plants is found in the fruit (60–66%; Winsor *et al.*, 1958). Deficiency symptoms have been found on mature fruiting plants when the leaves contained up to 2.4% K (Adams *et al.*, 1978b) and values below about 3.0% K may be regarded as sub-optimal for glasshouse tomato crops. The leaves of healthy plants contain 4.0% K (Ward, 1963), 5.5% K (Winsor, 1973), 4.4-5.5% K (Winsor and Massey, 1978) and 4.4–5.6% K (Adams *et al.*, 1978b). Higher concentrations (over 6% K) have been suggested for good fruit quality and flavour

(Adams *et al.*, 1978a; see Fig. 18). Sheldrake (1981) considered the optimum range to be 4.0–7.0% K, and values up to 7.5% K have been recorded (Winsor *et al.*, 1965b) for the middle foliage (15th leaf below top).

Widders and Lorenz (1982), working with field-grown determinate tomatoes, found that K concentrations in the laminae and petioles increased during early vegetative growth but declined after the onset of fruiting. Thus, averaged over four cultivars, the percentage of K in the laminae increased from 2.68% measured 50 days after planting to 3.22% at 80 days, declining progressively to 1.63% at 125 days. The content of the dried petioles declined from 6.02% at 60 days to 2.60% at 125 days.

The K content of tomato leaves is depressed by high levels of N application. This effect is illustrated in Table 14 from a 72-plot factorial study, the layout and early results from which were reported by Winsor *et al.* (1967). The quality of the fruit, judged both by uniformity of ripening and by absense of hollow ('puffy') tomatoes, improved progressively from K1 to K3 (P < 0.001).

Nitrogen application (kg ha^{-1})	Potassium application (kg ha^{-1})			Mean
	K1 (405)	K2 (810)	K3 (1620)	
N1 (238)	4.33	6.45	6.77	5.85
N2 (475)	3.73	5.32	6.33	5.13
N3 (950)	3.44	4.61	5.87	4.64
Mean	3.83	5.46	6.32	

Table 14. Effects of nitrogen and potassium application on the potassium content (% K in dry matter) of the leaves of soil-grown tomato plants. Each value in the body of the Table is the mean of analyses from eight plots (Winsor and Hart, unpublished data). LSD (P=0.05) N and K means 0.21 : N × K means 0.36.

Calcium

The Ca status of a solid substrate is rarely the factor which limits uptake of Ca by greenhouse crops, since lime (calcium carbonate) or dolomitic lime (magnesium and calcium carbonates) and P fertilisers (containing calcium phosphate and calcium sulphate) are generally incorporated before planting. In solution culture, adequate levels of this nutrient are usually maintained with calcium nitrate. Thus, depression of the Ca status of the plants and the incidence of deficiency symptoms usually result from the influence of other factors which impede either Ca uptake or its distribution within the plant. Conditions which commonly reduce uptake include the competitive effects of high concentrations of other cations e.g. K, Na, Mg or NH_4 in the substrate. Also, since Ca moves almost exclusively through the xylem, its uptake is affected by low root temperature and by restricted movement of water through the plant due to dryness or salinity in the substrate or to excessive humidity in the atmosphere.

Symptoms of calcium deficiency. Fisher (1935) found very rapid responses to omission of Ca in the sand culture, symptoms of deficiency developing within 3-5 days. The plants lost turgor, the upper stems became spotted with dead areas and the growing point died. The upper leaves, at first darker green than normal, soon yellowed at the edges and withered. Hewitt (1944) recorded yellow, orange and purple pigments at the centre of terminal leaflets; these spread to the lateral leaflets and the leaves eventually collapsed and died. The description given by Wallace (1951) included yellow, brown and purple tints in the terminal leaflets and restriction of growth. Under conditions of severe deficiency the growing point was scorched and the main stem died back. The illustration by Smilde and Roorda van Eysinga (1968) showed yellowing and scorching of the leaf margins together with some interveinal yellowing; the leaves were strongly curled downwards and inwards. The roots were poorly developed and brown in colour, confirming earlier observations by Fisher (1935) and Hewitt (1944). Death of the growing point and terminal leaflets was also illustrated by Bergmann (1976). The trusses failed to set and withered at the distal end. The fruit tended to ripen less rapidly at the blossom end (Hewitt, 1944). Severe BER develops on the fruit, particularly on the smaller ones at the end of the truss (Wallace, 1951); this topic is discussed more fully in a later section.

Hall (1977) noted symptoms of Ca deficiency on the youngest leaves of young tomato plants grown for 30 days in aerated solution culture at 1 mg l^{-1} Ca, whereas the lower leaves were unaffected. Maximum dry matter production occurred at the relatively low concentration of 4 mg l^{-1} Ca in solution, corresponding to 1.04% Ca in the leaves. It was suggested that high concentrations were adverse under conditions of short days and low light intensities owing to increased synthesis of oxalate at the expense of carbohydrates.

Responses to liming. Excessive amounts of lime are unfavourable to plant growth, since the high pH reduces the availability of other nutrients. For this reason, deficiencies of Fe, Mn and B frequently occur on calcareous soils. The adverse effect of liming on the yield of fruit is well known, but the response is variable. For example, Martens (1963) found that liming a sandy soil (pH 5.5) reduced the yield only slightly whilst Winsor et al. (1967) showed losses in yield due to liming which increased with the level of N applied; at the highest N level the yield was depressed by an average of 12% at pH 7.2. Similarly in peat, the yield of fruit declined progressively as the pH increased with liming from 4.5 to 6.7; reductions in yield at pH 6.7 ranged from 18% to 24% (Graves et al., 1978).

Heavy liming also affects fruit quality. Thus, Winsor and Long (1968) found an increase in fruit size in limed soil, accompanied by a reduction in the number of fruits formed. The proportion of irregularly shaped fruits was also increased by liming. Liming generally increased the proportion of unevenly ripened fruit from plants grown in soil (Winsor and Long, 1967) and in peat (8% at pH 5.5; 14% at pH 6.8: Adams, 1978b). The firmness of the fruit was, however, improved by liming (Shafshak and Winsor, 1964). Hamson (1952) had earlier demonstrated an increase in firmness with increasing Ca concentration in solution culture. The use of calcium chloride to increase the firmness of canned tomatoes has long been known (Kertesz, 1939).

Blossom-end rot. There is extensive evidence that the disorder known as blossom-end rot is closely associated with a deficiency of Ca in the fruit. The symptoms first appear as a water-soaked region round the stylar scar, turning very dark brown and forming a depressed area as the tissue loses water. The disorder can cause considerable losses of marketable fruit, both in field and protected crops. Premature ripening of those parts of the fruit adjacent to the brown tissue is highly characteristic of BER; under conditions favourable to this disorder the first fruits to ripen on the plant tend to be those affected by it.

Raleigh and Chucka (1944) reported that the incidence of BER on fruit from plants grown in solution culture decreased from 55% at 125 mg l^{-1} Ca to 3.4% and 2.4% at 500 and 1000 mg l^{-1} Ca respectively; a higher proportion of BER was found at 2000 mg l^{-1} Ca, but this was attributed to the very high Cl content of the solution used for this treatment. Chiu and Bould (1976) showed that the incidence of BER was influenced largely by Ca supplied directly from the roots during fruit development. Low Ca levels in solution culture before fruit development had little or no effect on the disorder. It thus appears that little Ca is translocated to the fruits from the leaves, making leaf analysis of limited value in diagnosing this order except under conditions of extreme deficiency. Fruit from plants grown throughout at 16 mg l^{-1} Ca showed 66% BER compared with none at 160 mg l^{-1}. The incidence of BER was reduced by spraying with 1% calcium chloride, and eliminated entirely by a combination of Ca sprays and injection of 2% calcium gluconate into the fruit. Calcium sprays e.g. 0.2% calcium nitrate, are widely used for glasshouse tomato crops in the U.K. whenever this disorder appears. Borkowski and Ostrzycka (1973) emphasized the importance of spraying the green fruits; spraying the leaves alone had no beneficial effect.

The incidence of BER is greatly increased by the use of high levels of NH_4-N. Barke and Menary (1971) reported 14.5% BER from plants grown in the field with ammonium sulphate (246 kg ha^{-1} every two weeks) compared with 3.6% in the untreated standard. Pill et al. (1978) grew tomato plants in sand culture at 70, 175 and 280 mg l^{-1} N, supplied either as NO_3-N or as NH_4-N. Plants grown with NH_4-N produced 35–64% BER as compared with none on plants grown with NO_3-N. Massey and Winsor (1980a) found increased incidence of BER with increasing proportions of NH_4-N in the total N supplied to young plants in nutrient film culture, accompanied by a depression in Ca and Mg content of the leaves. Raleigh and Chucka (1944) also demonstrated increased incidence of BER at high concentrations of K (1600 mg l^{-1}) and Mg (800 mg l^{-1}) in sand culture. The authors noted that the accompanying high levels of Cl might have contributed to this response. The decrease in Ca content (%) of the fruit at high K (41%) and high Mg (45%) nevertheless suggested that K:Ca and Mg:Ca antagonism occurred; see also Table 15.

Magnesium content of peat substrate (mg l^{-1} Mg)	Calcium content (% Ca in dry matter)	
	Leaves	Fruit
120	4.9	0.43
360	4.3	0.30
640	3.1	0.25

Table 15. Influence of magnesium level on the calcium content of tomato leaves and fruit (after Borkowski and Szwonek, 1979).

Evans and Troxler (1953) found slightly but significantly higher levels of Ca in normal tomato fruits (0.09%) than in those affected by BER (0.07%), these and subsequent values in this section being expressed on a dry matter basis. Wiersum (1966) stated that a Ca content of less than 0.08% in the fruit always gave rise to BER whereas, above 0.12%, all fruits were normal. Van Goor (1968) found 0.09% Ca in normal fruits and 0.03-0.04% in BER fruits.

Chiu and Bould (1976) induced BER by low Ca levels in solution culture and found 0.11% Ca in normal fruits compared with 0.03% in those affected by the disorder. Pill et al. (1978) reported 0.04% and 0.02% Ca respectively in normal and affected fruits from plants grown with NH_4-N in sand culture; both these values are low compared with 0.18−0.25% Ca found in fruits grown with NO_3-N, none of which showed BER.

As seen from these examples, many workers have found that the occurrence of BER is related to the Ca content of the fruit, being commonest at low levels. The actual values differ widely, however, with BER sometimes recorded at Ca levels higher than those found by other workers in normal fruit. The data of Ward (1973) are of particular interest in this connection. Ward found that, where BER was induced experimentally by low Ca levels, the affected fruits contained 0.02−0.03% Ca compared with 0.07−0.09% Ca in normal fruits from plants receiving adequate calcium. Ward also induced BER by subjecting the plants to moisture stress in the presence of an adequate supply of Ca; under these conditions the affected fruit contained 0.07% Ca, a value comparable with those sometimes found in normal fruits. Ward therefore suggested that there may be two physiological conditions resulting in identical symptoms. Moisture stress was shown by Shaykewich et al. (1971) to increase the incidence of BER, yet some BER occurred in all three water regimes tested, indicating that this factor, too, does not entirely control the incidence of the disorder. Other factors have also been shown to influence the proportion of fruits affected by BER. Thus Wiersum (1966) increased the incidence of BER to 48% by enclosing the fruit trusses in plastic bags compared with 7% from the untreated plants; the result was attributed to reduced transpiration from the fruits, linked with less movement of Ca in the transpiration stream. Bradfield and Guttridge (1984) showed that the incidence of blossom-end rot was increased by low humidity at night and by a high nutrient concentration in the root zone. It was suggested that a positive root pressure at night promotes transport of Ca into tissues and organs that have restricted transpiration.

Evans and Troxler (1953) noted that the incidence of BER was widely known to be associated with vigorous vegetative growth, and growth regulating chemicals have been shown to have marked effects on the occurrence of BER (Castro and Malavolta, 1976). For example, gibberellic acid doubled the incidence of BER whereas other growth regulators such as cyclocel (chlormequat chloride), a growth retardant, greatly reduced it.

The conditions influencing BER are thus complex and not fully understood, yet not necessarily contradictory. Calcium deficiency in the fruit, whether induced by inadequate supply to the roots or by limited transport to the fruit, could lead to increased cell permeability and breakdown; see, for example, Van Goor (1968). Thus Christianson and Foy (1979) concluded that 'the primary function of Ca in plants appears to be that of membrane stabilization'. Whilst water stress, shown by Carolus *et al.* (1965), Gerard and Hipp (1968) and Ward (1973) to induce BER, would affect the uptake and distribution of Ca in the system, it could also have a more direct effect on the disorder. Gerard and Hipp (1968) concluded that 'the high transpiration from small fruit and slow movement of water in the fruit parts create a condition whereby water loss from fruit can exceed water intake. Under high evaporative conditions, transpiration losses from small fruit probably cause collapse of the sensitive, unstable tissues and result in blossom-end rot'. In view of the known effect of Ca in stabilising cell membranes it seems likely that the interplay of Ca status with water stress could account for the rather wide range of Ca levels in the fruit at which BER occurs.

Calcium content of the leaves. The Ca content of the youngest leaves is relatively low (1–2% Ca), whereas it accumulates in the oldest leaves (6–7%; Ward, 1964). Heavy applications of N, K or Mg depress the Ca content of the plants and the fruit (Raleigh and Chucka, 1944). The form of response to increasing levels of K is shown at two levels of liming (pH 6.1 and 7.2) in Figure 19. The influence of Mg level on the Ca content of the the leaves and fruit is shown in Table 15, p.59. Calcium uptake is influenced greatly by the form of N supplied: for example, the content of young plants receiving only NH_4–N was depressed by over 80% compared with plants supplied with NO_3–N only (Wilcox *et al.*, 1973; see also Table 12, p.53).

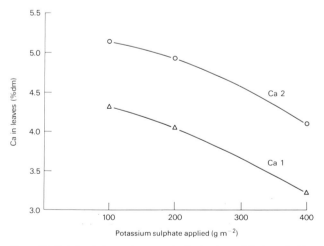

Figure 19   Relation between the amounts of potassium sulphate applied to the soil before planting and the calcium content of tomato leaves (Adatia, unpublished data).

Symptoms of deficiency are generally associated with concentrations of less than 1% Ca in the leaves (Ward, 1963; MacLean *et al.*, 1968); Wallace (1951) reported

0.56% Ca and Bergmann (1976) 0.71% Ca in leaves affected by severe deficiency. The values associated with healthy growth include 1.5% Ca (Ward, 1963), 1.3-1.7% Ca (MacLean *et al.*, 1968), 2.5% Ca (Winsor, 1973), 3.3% Ca (Maher, 1976) and 1.0–5.0% Ca (Sheldrake, 1981). Humphries and Devonald (1977) found a steep gradient in the Ca content of tomato leaves from 4.1% Ca in the bottom leaf to 0.9% Ca at the top (8th leaf); the young plants had been grown in solution culture at 80 mg l^{-1} Ca. Drake and White (1961) showed that both Ca uptake (mg per plant) from a limed soil and percentage in the young tomato plants (65 days old) increased with the rate of application of N as ammonium nitrate. The content of the whole tops of the plants ranged from 0.82% Ca at low N (168 kg ha^{-1}) to 1.78% Ca at higher N (504 kg ha^{-1}). The total Ca content of the foliage is not a very reliable indicator of the Ca status of the fruit, since there is virtually no movement from the leaves to the fruit (Chiu and Bould, 1976).

Magnesium

Symptoms of deficiency. Magnesium deficiency is perhaps the commonest nutritional disorder of glasshouse tomatoes, and is doubtless accentuated by the high levels of K necessary for production of high quality fruit. Interveinal chlorosis and necrosis of tomato leaves have long been associated with this deficiency; for early descriptions and references see chapter by McMurtrey in Kitchen (1948). Cromwell and Hunter (1942) reported widespread chlorosis of tomato foliage in the west of Scotland and showed this to be associated with low Mg levels. The vascular system and adjacent zones of the laminae, and also the leaf margins, remained green while the interveinal areas became yellow or greenish-yellow. Hewitt (1944) described interveinal yellowing spreading from the leaf margins, starting on the older leaves and spreading up the plant; the old leaves became brown and withered. Wallace (1951) also noted interveinal chlorosis and necrosis, progressing rapidly from the older to the younger leaves and accelerated by heavy fruiting. The margins of the leaves may remain green. Woolley and Broyer (1957) recorded interveinal chlorosis but also noted sunken necrotic spots which appeared shiny from the back of the leaf.

The symptoms generally appear first on the middle leaves when the plants are carrying a heavy load of fruit. Studies in nutrient film culture (Winsor, Hurd and Price, 1979) have shown that when the maximum load of fruit is developing on the plants i.e shortly before the first harvest, root growth ceases and some of the older roots begin to die off. It seems likely, therefore, that the incidence of deficiency at this stage in the crop may be related to the health and development of the root system.

Responses to magnesium. Jones *et al.* (1945) concluded that dressings of magnesium suphate equivalent to 227 kg ha^{-1} Mg were necessary to obtain satisfactory commercial control of deficiency symptoms. Spraying the plants with magnesium sulphate (2.5% solution of $MgSO_4.H_2O$ plus a wetting agent) was very effective, whereas applying this solution to the soil (113 kg ha^{-1} Mg) was much less so. As with most foliar sprays, there is some risk of leaf scorch, and lower concentrations (1–2% $MgSO_4.7H_2O$) are now favoured.

Mild Mg deficiency has little effect on fruiting, but moderate and severe deficiencies may cause appreciable loss of yield. For example, in one long-term trial the plants not supplied with Mg showed symptoms of deficiency for four consecutive years but an appreciable depression in yield (12%) was found only in one year which was particularly sunny (Winsor et al., 1965a). In a later trial the yield was not affected significantly by Mg level until the ninth year, when the deficiency depressed the yield by 11% at the highest level of K. With a late-planted crop in the tenth year of the trial, the loss in yield due to Mg deficiency increased with the levels of N or K applied and amounted to 29% where the plants received high levels of both nutrients (Adatia and Winsor, 1971; see also Table 16). Similarly, the loss in yield due to the deficiency was greater in heavily limed soil (18%) than in unlimed soil (6%). The results of these trials, in which the plants were grown in the border soil, suggest that an appreciable loss in yield occurs only when the Mg content of the substrate has declined to a very low level.

Potassium level	Yield (kg) Magnesium level		Difference (Mg 2 − Mg 1)	Difference as % of yield at Mg 2
	Mg 1	Mg 2		
K1	3.67	4.04	0.37	9.2
K2	3.89	5.00	1.11	22.2
K3	3.07	4.68	1.6l	34.4

Table 16. Effect of magnesium deficiency on the yield of tomatoes (kg per plant), and the interaction between magnesium and potassium for plants grown at a high level of applied nitrogen (after Winsor and Adams, 1968).

Yield responses have also been found in tomatoes grown in soilless substrates. For example, in the third year of a trial in peat beds, Mg deficiency resulted in a depression in yield of 12–14% at the higher levels of K (Adams et al., 1978b).

Dressings of magnesium sulphate increase the salinity of the soil and have been shown to reduce fruit size (4%) and the proportion of irregularly shaped fruits (10%; Winsor and Long, 1968); the percentage of unevenly ripened fruit was also decreased (17%; Winsor and Long, 1967). The application of magnesium sulphate had little effect, however, on the quality of tomatoes grown in peat (Adams et al., 1978a).

Magnesium content of the leaves. Leaves from healthy plants usually contain more than 0.4% Mg; Wallace (1951) recorded 0.48% Mg, Ward (1964) suggested 0.4-0.6% Mg and Winsor (1973) quoted 0.5% Mg, whereas Smilde and Roorda van Eysinga (1968) gave somewhat higher values (0.6–0.9% Mg). Still higher values were found by Markus and Steekel (1980) in crops grown commercially in a peat-vermiculite substrate. Nine out of the thirty leaf samples contained 1.0% Mg or more, the overall range being 0.59–1.80% Mg. In plants adequately supplied with Mg the content of this nutrient was greater in the lower (15th) than in the higher (5th) leaves (P < 0.01; Winsor et al., 1965b). The opposite trend can occur in deficient plants, however, matching the distribution of chlorosis.

The Mg content of tomato leaves from plants showing varing degrees of chlorosis is given in Table 17, based on experiments by Jones et al. (1945); 0.3% Mg in the leaves was considered to be the threshold between sufficiency and deficiency. Using the fifteenth leaf below the head of the plant, Winsor et al. (1965b) found an approximately linear relationship (r = −0.94^{xxx}) between the severity of the symptoms and the content of the leaves up to about 0.4% Mg (Fig. 20). Severe deficiency is associated with contents of less than about 0.2% Mg; for example, 0.13% Mg (Wallace, 1951), 0.15% Mg (Ward, 1964), 0.17–0.26 Mg (MacLean et al., 1968) and 0.13–0.19% Mg (Adatia and Winsor, 1971). Johannesson (1951) reported values down to 0.06% Mg in leaves with severe chlorosis. Significant losses in yield correspond to contents of 0.25% or less e.g. 0.22% Mg (MacLean et al., 1968), 0.23 Mg (Adatia and Winsor, 1971) and 0.25% Mg (Adams et al., 1978b). Maximum yield was achieved when the leaves contained 0.31–0.40% Mg (Adams et al., 1978b).

	Mg applied** (kg ha^{-1})	K applied	% Mg in leaf	Symptoms
Site 1 (June*)	136	0	0.19	Slight
	0	0	0.14	Medium
	136	+	0.16	Slight/Medium
	0	+	0.12	Medium
Site 2 (July*)	136	0	0.28	Slight
	0	0	0.14	Severe
	136	+	0.28	Slight
	0	+	0.14	Severe
Site 3 (December*)	136	0	0.30	Trace
	0	0	0.19	Medium
	136	+	0.26	Slight
	0	+	0.19	Medium

Table 17. Incidence of deficiency symptoms and % Mg in tomato leaves at three glasshouse sites (data from Jones et al., 1945).
* Date of visual assessment and leaf sampling.
** Applied as calcined kieserite ($MgSO_4.H_2O$).

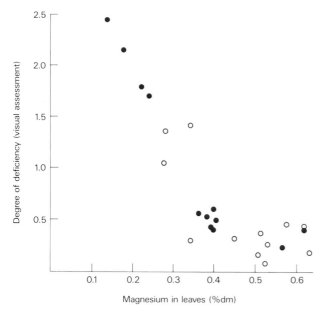

Figure 20 Relation between the magnesium content of tomato leaves (15th below the top) and the severity of magnesium deficiency symptoms, assessed visually from yellowing of the leaves (0, Low potassium; ●, High potassium; after Winsor *et al.*, 1965b).

Magnesium sulphate applied (kg ha^{-1})$^+$	Potassium sulphate applied (kg ha^{-1})$^+$			LSD (P=0.05)
	888	1776	3570	
Nil	0.33	0.31	0.26	0.04
2034	0.97	0.73	0.61	0.04

Table 18. Effect of the rate of application of potassium sulphate on the magnesium content (% Mg) of tomato leaves (after Adatia and Winsor, 1971).
$^+$Total amount applied both before planting and during cropping.

The Mg content of the leaves was depressed by high rates of application of K (Walsh and Clarke, 1945a; see also Table 18). Heavy liming also depressed the Mg content from 0.33% to 0.26% Mg (Adatia and Winsor, 1971). In the limed soil, heavy applications of superphosphate (7.9% P) further reduced the content from 0.19% to 0.13% Mg where high levels of N and K had also been applied. Plants receiving only light dressings of N and K contained more than 1% Mg in the leaves.

Iron

Deficiency symptoms. The uppermost leaves of Fe deficient plants become pale yellow-green in colour with central interveinal mottling, commencing at the base of the compound leaves and leaflets. Marginal and interveinal scorching develop on the young expanded leaves, with slight purpling of the petioles (Hewitt, 1945). In addition

to chlorosis of the terminal leaves, the top of the main stem becomes pale yellow-green in colour and growth is restricted (Wallace, 1951). Woolley and Broyer (1957) noted that the smallest veins do not remain green and that the brown patches which occur on some leaves are not associated with any particular part of the laminae. Good illustrations of the deficiency have been published by Wallace (1951) and by Roorda van Eysinga and Smilde (1981). The latter authors noted that, at first, even the smallest veins remain green to produce a fine reticular pattern on a yellow background; eventually, however, the whole leaf becomes chlorotic.

When induced in solution culture, Fe deficiency greatly reduced plant growth; dry matter production was depressed by over 75% and the yield of fruit by over 70% (Lyon *et al.*, 1943). However, this deficiency rarely occurs commercially as a result of a lack of Fe in the substrate. It is induced most commonly by conditions such as poor aeration, often due to overwatering, by root disorders or by a high pH (< 7.0) in the substrate. The Fe content of the foliage may be depressed by high levels of other heavy metals e.g. Ca, Mn and Zn. Thus, translocation of Fe from the roots to the leaves decreased as the supply of Mn increased, and Fe accumulated in the roots (Riekels and Gulun, 1967).

Iron content of the leaves. The Fe content of the leaves associated with deficiency varies from 74 ug g^{-1} (Lyon *et al.*, 1943), or less than 88 ug g^{-1} (Twyman, 1951), to 210 ug g^{-1} (Walsh and Clarke, 1945b). Stanton (1966) reported fluctuations in the content of leaves between seasons, the median ranges being 120−190 ug g^{-1} and 70−80 μg g^{-1} in successive years; treatment of the soil with Fe chelate gave slightly improved colour in both the young and older leaves. Healthy leaves have been reported to contain 269 μg g^{-1} Fe (Wallace, 1951), 155−819 μg g^{-1} (Smilde and Roorda van Eysinga, 1968) and 100−250 μg g^{-1} (Sheldrake, 1981). The Fe content of tomato leaves was depressed progressively from 84 μg g^{-1} to 61 μg g^{-1} by liming peat plots which had been treated with a fritted micronutrient mixture from pH 4.3 to pH 6.3. Where the micronutrients were omitted the leaves of plants grown at the highest pH contained only 27 μg g^{-1} Fe (Graves *et al.*, 1978). The diversity of the values associated with either normal or deficient plants suggests that the *total* content of the leaves does not necessarily reflect the amount of physiologically active Fe contained in them; the results cannot always be interpreted with certainty. There is scope for the further development of tests assessing the content of 'available' Fe in plant tissues (see, for instance, De Kock, 1983). Enzymic tests involving Fe-containing enzymes e.g. peroxidase (see Vol. 1, p.131), have shown some differentiation between deficient and normal tomato leaves (Besford, 1975), but such tests cannot as yet be regarded as established procedures for identifying Fe deficiency in tomatoes.

Manganese

Symptoms of deficiency and toxicity. Manganese deficiency may occur on calcareous or heavily limed soils. The mature leaves are reduced in size, and an interveinal mottle develops which is pale green at first but later becomes yellow; the veins remain dark green. Small brown spots develop in the yellow areas, beginning near the base of the leaflets. The root system is much reduced in size, being shorter and finer than normal with some browning of the tips (Hewitt, 1944). Wallace (1951) reported that the mottling was particularly noticeable in the upper part of the plant but subsequently spread to all the leaves. The chlorosis was not so intense as for Fe deficiency, but the mottled areas could become necrotic. Illustrations of the deficiency symptoms were published by Wallace (1951), Bergmann (1976) and Roorda van Eysinga and Smilde (1981).

Excessive amounts of Mn lead to the formation of brown lesions on the petioles and stems, especially near the nodes, and the leaves wither (Wallace, 1951). Lamb (1961) described a slight but distinct interveinal chlorosis with down-curling of the leaflets. Large necrotic patches developed very rapidly and soon affected whole leaflets. The tips of the sepals also withered and died. Smilde and Roorda van Eysinga (1968) show the typical dark deposit which forms along veins, accompanied by interveinal chlorosis.

Manganese content of the leaves. When Mn deficiency was induced in solution culture, growth was depressed markedly; the yield of fruit was reduced by 74% and the Mn content of the upper leaves decreased from 70 μg g^{-1} to 6 μg g^{-1} Mn (Lyon *et al.*, 1943). The Mn content of the leaves of deficient plants showed virtually no gradient down the plant whereas, when included in the nutrient solutions, the values increased progressively from 70 μg g^{-1} Mn in the top third of the plant to 248 μg g^{-1} in the middle portion and 398 μg g^{-1} in the lower leaves (Lyon *et al.*, 1943). Ward (1977) found some increase in Mn content of deficient tomato plants from the uppermost (17th) leaf laminae (1 μg g^{-1} Mn) to the bottom (2nd) leaves (24 μg g^{-1} Mn). The corresponding values for plants receiving adequate Mn were 54 μg g^{-1} Mn at the top (16th leaf) to 87 μg g^{-1} at the bottom. Values for the fifth leaf from the top, widely adopted for diagnostic tests, were 15 and 58 μg g^{-1} Mn in deficient and normal plants respectively. Wallace (1951) found only 7 μg g^{-1} Mn in deficient leaves. Symptoms of the deficiency were found in young plants containing 23 μg g^{-1} Mn (Wallihan and Bradford, 1977). In more mature plants, deficiency symptoms were associated with leaves containing 25 μg g^{-1} Mn or less, although this concentration was also found in some leaves without symptoms (Ward, 1977).

Other Mn values reported for the leaves of healthy tomato plants include 46 μg g^{-1} (Wallace, 1951), 120-150 μg g^{-1} (Stanton, 1966; median range) and 27-239 μg g^{-1} (Smilde and Roorda van Eysinga, 1968; from their own research). Roorda van Eysinga and Smilde (1981) quoted a general range for healthy plants of 55−384 μg g^{-1} Mn, and Sheldrake (1981) regarded 40−300 μg g^{-1} Mn as optimal.

Excessive amounts of available Mn may be released in acid or waterlogged soils and can cause toxicity. However, in protected crops this disorder occurs most frequently as a result of partial steam-sterilisation of the soil; Davies (1957) found that the leaves of affected plants contained up to 4900 μg g^{-1} Mn. Menary and Kruger (1966) noted that the dry weight of tomato seedlings declined when the tissue contained more than 1000 μg g^{-1} Mn.

Dennis (1968) found that severe toxicity stopped the growth of seedlings four weeks after they were planted in a slightly acidic steamed compost (pH 5.4). The cotyledons became yellow and dropped; the yellowing spread to the true leaves, and brown blotches formed on the petioles and stems. The dried tissue of the whole plants contained 5240 μg g^{-1} Mn. Liming the same soil to pH 6.5 in a subsequent trial reduced the manganese content of the tissues to 670 μg g^{-1}. Davies (1957) showed that heavy applications of superphosphate (7.9% P) to an acid soil (pH 5.3) reduced Mn uptake. The most effective control of Mn toxicity was obtained by applying both lime and superphosphate (Table 19); the combination is particularly effective since superphosphate itself is acidic. Lamb (1961) found that heavy dressing of superphosphate reduced the incidence of Mn toxicity symptoms and reduced the content of leaves from the middle of the plants from 4484 μg g^{-1} to 2260 μg g^{-1} Mn.

Soil treatment	Soil pH	Manganese content (μg g^{-1})	
		Soil [+]	Plants
Unsteamed	5.3	62	2900
Steamed	5.1	84	4900
Steamed plus 0.5% superphosphate	4.8	50	3600
Steamed plus 0.5% lime	7.0	31	1300
Steamed plus 0.5% superphosphate plus 0.5% lime	6.8	32	800

Table 19. The effect of steam-sterilisation and additions of lime and superphosphate on the pH and available manganese content of an acid field soil of low phosphate content and on the manganese content of tomato seedlings (after Davies, 1957).
[+] Water-soluble plus exchangeable manganese.

In nutrient film culture, the pH of the recirculating nutrient solution was found to have a much greater effect on the uptake of Mn than the concentration of this nutrient in the solution. Manganese concentrations in the young leaves within the range 137 μg g^{-1} to 291 μg g^{-1} had no effect on yield (Sonneveld and Voogt, 1980).

Ward (1977) found that the Mn content of plants affected by toxicity increased from 668 μg g^{-1} Mn in the top leaves to 5800 μg g^{-1} in the lowest leaves. The exact value above which toxicity is considered to occur is not easily defined, but 500 μg g^{-1} Mn in the youngest leaves and 900–1000 μg g^{-1} in the older leaves have been suggested (Ward, 1977). Milikan (1951) published paired photographs and radio-autographs of tomato leaves affected by Mn toxicity. Necrotic spotting at first coincided with high Mn levels in the interveinal areas, but the Mn later became concentrated in the veins.

Copper

Copper deficiency is comparatively rare in greenhouse tomato crops. It occurs mainly on soils with a very high content of organic matter or on soilless media such as peat, where the formation of stable complexes between humified organic matter and Cu reduces the availability of this nutrient to the plants (Sanders and Bloomfield, 1980). Severe deficiency occurred commercially in the U.K. where tomato plants were grown in peat modules from which Cu had inadvertently been omitted.

Symptoms and growth responses. Symptoms of Cu deficiency have been induced in solution culture (Sommer, 1931; Arnon and Stout, 1939a; Bailey and McHargue, 1943); growth was much restricted, the leaf margins curled upwards and inwards and became severely scorched, and flowering was delayed or suppressed. Very poor root development was also noted by Arnon and Stout (1939a). The first symptoms recorded by Lyon et al. (1943) were an overall grey-green colour followed by chlorosis of the lower leaves. The chlorotic leaves later became bronzed and finally brown, with necrosis of the margins and black veins. Brown woody spots developed on the stems and petioles. These symptoms may not all be present, however, when the deficiency is less severe. For example, the margins and tips of the leaves wilted and, on the old leaves, rolled up stiffly (Woolley and Broyer, 1957). The leaves of young flowering plants were described as blue-green in colour (Wallihan and Bradford, 1977), and dark green with limp margins and tips (Hewitt and Watson, 1980). The first symptom on more mature plants (10-12 weeks after planting) was wilting of the youngest leaves followed by yellowing of the oldest leaves (Adams et al., 1978c). Illustrations have been published by Smilde and Roorda van Eysinga (1968), and by Hewitt and Watson (1980).

Copper deficiency has been reported on a highly calcareous soil (pH < 8.5); the plants responded immediately when copper sulphate was applied to the soil (Lamb and Conroy, 1962). Copper deficiency reduced the yield of fruit from crops grown in peat by 11% (Adams et al., 1978c) and by 25% (Adams, 1978a). Whilst Adams and Winsor (1974) found no significant overall yield response to Cu in a glasshouse soil at pH 7.2, there was

some indication (P < 0.1) of a consistent B–Cu interaction in the data for yield, number of fruit and uniformity of ripening. Thus added Cu increased the yield from B deficient plants but not from plants adequately supplied with B. In this connection it has been noted that phenols accumulate in both B deficient and Cu deficient plants (Brown, 1979).

In solution culture, growth was very severely restricted by Cu deficiency; very few fruit were formed and both the fresh and dry weights of the plants were less than 10% of the weight of healthy plants (Lyon et al., 1943). On a calcareous soil, fruit on the upper trusses of deficient plants failed to swell, and both setting of the flowers and swelling of the fruit were improved by dressings of copper sulphate (Lamb and Conroy, 1962).

Copper content of the leaves. The leaves of deficient plants contained 4–5 μg g^{-1} Cu compared with 14–15 μg g^{-1} in healthy plants (Lamb and Conroy, 1962). In peat, heavy liming depressed the content of the leaves of plants adequately supplied with micronutrients from 12–16 μg g^{-1} Cu to 3–5 μg g^{-1} (Graves et al., 1978; see Table 20). Smilde (1972) reported 1.9–2.6 μg g^{-1} Cu in the leaves of

CaCO$_3$ (kg m^{-3})	pH	Frit 253A (g m^{-3})			
		Nil	200	500	Mean
1.75	4.3	3.4	12.3	15.8	10.5
3.0	4.9	2.9	8.9	14.1	8.6
4.5	5.4	1.5	4.9	8.3	4.9
8.0	6.3	2.6	3.2	4.6	3.4
Mean		2.6	7.3	10.7	

Table 20. The effects of a micronutrient mixture (Frit 253A) and liming on the copper content (μg g^{-1} B Cu) of leaves from tomato plants grown in peat (after Graves et al. 1978)

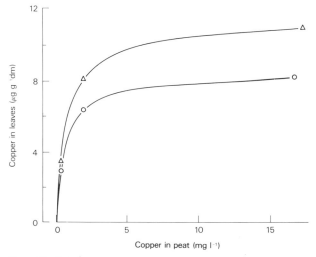

Figure 21 Relation between the copper contents of tomato leaves and of the peat substrate, using 0.05 M (NH$_4$)$_2$EDTA as extractant (after Adams et al., 1978c).

deficient tomato plants grown in peat. Adams *et al.* (1978c) found $3-4$ µg g^{-1} Cu in deficient plants compared with some $5-10$ µg g^{-1} in plants without symptoms. Sheldrake (1981) quoted an optimum range of $5-25$ µg g^{-1} Cu. The Cu content of the leaves is shown in Figure 21 in relation to that of a peat substrate at two levels of liming.

Zinc

Symptoms of zinc deficiency. Zinc deficiency rarely occurs in commercial glasshouse crops, but has been studied in solution culture. Lyon *et al.* (1943) observed interveinal chlorosis of the lower leaves, this condition spreading upwards on the plant. The lower leaves ultimately turned brown. Vegetative growth was restricted, and fruit production was reduced by 56%. Severe deficiency limited the growth of tomato plants in sand culture and caused the leaflets to curl downwards and sometimes completely round (Hewitt *et al.*, 1954). Cell contents diffused out to the exterior of the leaves, suggesting deterioration of the cell membranes. Smilde and Roorda van Eysinga (1968) found little chlorosis but reported brown spots on and between the veins of the leaflets and also on the petiolules (leaflet petioles). The petioles curled downwards, giving the leaves a strongly coiled appearance; see also Lingle *et al.* (1958) for field grown tomatoes. Zinc deficiency was illustrated by Hewitt and Watson (1980), who described chlorosis and twisting of the leaves. Short lengths of necrotic tissue developed along the petiolules near the base of the leaflets.

Zinc content of the leaves. The lower leaves of deficient plants were found by Lyon *et al.* (1943) to contain 17 µg g^{-1} Zn compared with 30 µg g^{-1} in healthy plants. The upper leaves were less affected by deficiency, the corresponding values being 23 µg g^{-1} and 28 µg g^{-1} Zn respectively. Lingle *et al.* (1958) reported $6.0-8.7$ µg g^{-1} in the leaves of deficient tomato plants grown in the field. Young plants showing deficiency contained 23 µg g^{-1} Zn (Wallihan and Bradford, 1977). Stanton (1966) found median ranges of $20-40$ µg g^{-1} and $25-40$ µg g^{-1} Zn in the leaves of healthy tomato plants during two successive years of glasshouse trials in New Zealand. The Zn content of the tissue was not increased by application of zinc sulphate ($41-112$ kg ha^{-1}) to the soil, though some slight indication of improved fruit quality was suggested. Smilde and Roorda van Eysinga (1968) found $48-66$ µg g^{-1} Zn and $201-458$ µg g^{-1} Zn in the leaves of healthy tomato plants grown on marine and river clay soils respectively.

The uptake of Zn may be depressed by high concentrations of other heavy metals or of P in the substrate. For example, increasing the level of Mn depressed the content of the leaves of the oldest plants (10 weeks) tested from 16 µg g^{-1} Zn to 10 µg g^{-1} Zn (Ward, 1977). Application of calcium phosphate (46% P_2O_5) to field-grown tomatoes in Texas induced severe symptoms of Zn deficiency (Burleson *et al.*, 1961); the content of the youngest fully developed leaves was decreased from 24.6 to 16.6 µg g^{-1} Zn (P < 0.001).

Zinc toxicity. Application of zinc sulphate in solution at rates of 0.33 and 0.65 g Zn per pot every second day to tomato plants in a loam:peat:sand substrate reduced growth (fresh weight) by 33% and 44% respectively (Nicholas, 1951). Forster (1951) reported a brown veinal necrosis and uniform chlorosis of the older leaves; veins near the margin of the leaves frequently showed pink or purple colours, especially on the dorsal surface. Smilde and Roorda van Eysinga (1968) noted spindly stunted growth. The younger leaves remained small with interveinal yellowing. Purpling developed on the undersides of the leaflets and the veins became reddish-brown. Leaf concentrations above about 327 µg g^{-1} Zn would be regarded as toxic (Roorda van Eysinga and Smilde, 1981). Data compiled by Chapman (1966) include a range of 526-1489 µg g^{-1} Zn in tomato leaves affected by Zn toxicity.

Chlorosis and a brown necrosis are sometimes seen in the foliage of glasshouse tomato plants directly affected by the dripping of water condensed on galvanised iron gutters.

Boron

Symptoms and growth responses. The symptoms of B deficiency vary considerably, those observed in sand or solution culture being generally the most severe and colourful. Thus Hester (1938), working with soil-grown plants, concluded that "Boron deficiency symptoms have been reported on the tomato before....but the symptoms were entirely different". Symptoms noted by Hester included yellowing of the tips of the leaflets of the oldest leaves, with prominent pink veins. The yellow area, though small at first, gradually enlarged and spread to other leaflets. Similar symptoms were illustrated in colour by Smilde and Roorda van Eysinga (1968) and by Winsor (1969). Yellowing of the tips of the lower leaves and brittleness of the leaflets and petiolules were characteristic symptoms of the deficiency on mature plants grown in soil (Adams and Winsor, 1974). Later the yellow leaflet tips dried out into a brown area which was associated with the main vein and surrounding tissues rather than with the leaf margin. Under conditions of severe deficiency a wide range of symptoms was found on the upper leaves, the most widespread of these being an orange-brown colour along the margins. Some leaves were pale cream in colour with purple margins, the leaflets being curled downwards, and in extreme cases the growing points died. Roorda van Eysinga and Smilde (1981) noted that, in commercial crops, the most striking symptom was yellow-orange discolouration of the leaves, particularly at the top of the plant.

In sand culture, the margins of the upper leaves became yellow at the base; reddish and purple tints developed and the veins were purple on the underside of the leaves. The petioles of the leaflets were brittle and the root system was greatly reduced in size, with browning and gelatinisation of the root tips, particularly of the youngest roots (Hewitt, 1944); for earlier descriptions see Johnston and Dore (1928); Johnston and Fisher (1930). With a severe deficiency the stems were short, thickened and stiff, the growing points died, and the leaves became highly coloured with yellow, brown and purple areas. With subsequent growth of the laterals, the plants became bushy and the leaves became distorted. Corky pitted areas developed on the fruit and ripening was uneven (Wallace, 1951). The first indication of B deficiency observed by Messing (1957) was chlorosis of the youngest leaves, starting at the base of each leaflet. The leaflets were distorted, curled, less smooth and thicker than on normal plants, and the leaves tended to be more compact as the petioles were shortened. The stems and petioles were brittle, leading to difficulties when training the plants. Illustrations published by Messing (1957) show that, under conditions of severe B deficiency, the fruits frequently failed to set or to develop, whilst those that matured had corky areas around the shoulders, quite close to the calyx. Brittleness of the tissues and distorted growth were included as diagnostic features of the deficiency by Wooley and Broyer (1957). Under some conditions, the walls may split to reveal the contents of the locules (Maynard et al., 1959). Distortion of the young leaves and death of the growing point were illustrated by Hewitt and Watson (1980). Chlorosis and browning of the leaves were shown by Roorda van Eysinga and Smilde (1981), together with death of the smallest leaflets and growing points.

Symptoms of deficiency are accentuated by liming and by heavy applications of P (Adams and Winsor, 1974) or K (Reeve and Shive, 1944; Adams and Winsor, 1974).

Boron deficiency had a marked effect on growth and fruit production when induced in sand culture (Johnston and Fisher, 1930; Messing, 1957), and the deficiency is often found in commercial crops e.g. Lamb and Conroy, 1962, particularly in the early stages of growth when watering is restricted. The addition of B to a soil limed to pH 7.2 increased the yield by 18% and also improved the fruit shape and uniformity of ripening; the response was greatest where high levels of N and P had been applied (Adams and Winsor, 1974). Boron deficiency was severe in heavily limed peat (pH 7.0); the yield was depressed by 50% (Adams, 1978a) and the proportion of unevenly ripened fruit increased from 15% to 45% (Adams, 1978b).

Boron content of the leaves. The leaves of healthy tomato plants have been reported to contain 46 μg g^{-1} B (Wallace, 1951) and 24-125 μg g^{-1} (quoted by Smilde and Roorda van Eysinga, 1968, from published sources). MacLean et al. (1968) found 38-60 μg g^{-1} B in healthy leaves from a range of commercial nurseries, and Sheldrake (1981) quoted an optimum range of 35-100 μg g^{-1} B. Messing and Winsor (1957) showed that the content of the leaves generally increases with age. For example, analyses of composite samples of laminae at each successive position below the top of the plants showed a progressive increase from 39 μg g^{-1} B in the second leaf to 113 μg g^{-1} B in the sixteenth leaf. The positional effect was greatest at the top of the plants, making the selection of a well-defined sampling position particularly important. Thus, in the example quoted, leaves one above and one below the customary fifth leaf from the top gave values 20% lower and 18% higher respectively. The same authors also studied the B content of the fifth leaf of a quasi-commercial tomato crop throughout the season; the values increased from 40 μg g^{-1} B in the laminae during early May to 77 μg g^{-1} B in late August and early September, possibly reflecting less vigorous vegetative growth as the plants become older. Majewski and Majewska (1953) found that deficiency symptoms occurred when the foliage contained less than 25 μg g^{-1} B, and Gupta (1983) reported deficiency in young tomato plants below 12 μg g^{-1} B. A range of 7−30 μg g^{-1} B was reported for the leaves of deficient plants (Smilde and Roorda van Eysinga, 1968) and is in good agreement with observations suggesting that 9, 19 and 27 μg g^{-1} B correspond to very severe, severe and moderate degrees of the deficiency (Adams and Winsor, 1974). Severe symptoms of the deficiency were associated with 7−15 μg g^{-1} B in the leaves of plants grown in peat without added micronutrients (Graves et al., 1978; see Table 21). The relationship between the severity of the symptoms and the B content of the leaves is shown for a soil-grown crop is Figure 22.

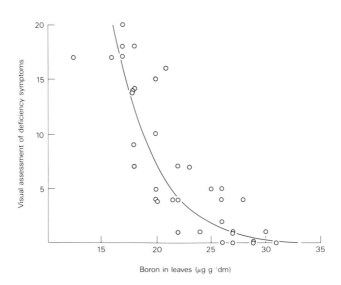

Figure 22 Relation between the severity of boron-deficiency symptoms and the boron content of leaves from tomato plants grown in soil without supplementary boron (after Adams and Winsor, 1974).

CaCO$_3$ (kg m^{-3})	pH	Frit 253A (g m^{-3})			
		Nil	200	500	Mean
1.75	4.3	15	33	34	27
3.0	4.9	13	30	33	26
4.5	5.4	8	26	32	22
8.0	6.3	7	10	21	13
Mean		6	25	30	

Table 21. The effect of a micronutrient mixture (Frit 253A) and liming on the boron content (μg g^{-1} B) of leaves from tomato plants grown in peat (after Graves *et al.* 1978).

Excessive levels of B result in scorching of the leaf margins, starting on the lower leaves and progressing upwards; see Bergmann (1976); Roorda van Eysinga and Smilde (1981). MacKay *et al.* (1962) reported symptoms of toxicity when the leaves contained 125 μg g^{-1} B. Wallihan *et al.* (1978) found toxicity symptoms at above 170 μg g^{-1} B in the leaves whilst Gupta (1983) found toxicity in young tomato plants above 172 μg g^{-1} B. Majewski and Majewska (1953) reported toxicity at 200 μg g^{-1} B. Very high concentrations may be found when the toxicity is severe. Thus, 300–500 μg g^{-1} and 450–900 μg g^{-1} B were found in the young and old leaves respectively (Brennan and Shive, 1948), whilst Oertli and Kohl (1961) found 2280–3130 μg g^{-1} and 3900–5150 μg g^{-1} B in chlorotic and necrotic tissue respectively. Bergmann (1976) reported 200 μg g^{-1} B in young leaves; the central parts of older leaves from the same plant contained 480 μg g^{-1} B whilst 1000 μg g^{-1} B were found in their margins.

Molybdenum

Molybdenum deficiency occurs on acid soils and peats, and is usually alleviated more readily by liming the substrate or by spraying the plants with a solution of a Mo salt e.g. 0.05% sodium molybdate, rather than by the application of Mo to the substrate, though the latter treatment can also be effective. Symptoms of the deficiency, induced in solution culture, appeared first as a distinct mottling of the lower leaves (Arnon and Stout, 1939b). Later the margins became scorched and inrolled. Most of the flowers dropped off before any fruits were set. Plants sprayed with a solution containing only 0.05 mg l^{-1} Mo resumed normal growth, and symptoms of the deficiency disappeared. Wallace (1951) described the leaflets of Mo deficient plants grown in sand culture as somewhat chlorotic, strongly incurled and dying back from the tips. A similar description was given by Hewitt and Bolle-Jones (1952), who noted that the leaf symptoms eventually spread to the youngest leaves. The old leaves withered and dropped prematurely and the growing point and whole plant ultimately died. The pale green to yellow interveinal chlorosis was illustrated by Smilde and Roorda van Eysinga (1968), who also noted that the smallest veins do not remain green. Woolley and Broyer (1957) stressed that bright colouration was not common in the chlorotic leaves of Mo deficient tomato plants.

Molybdenum content of the leaves. Healthy plants usually contain more than 0.3 μg g^{-1} Mo; the range of values reported includes 2.5 μg g^{-1} Mo in young leaves and 4.8 μg g^{-1} in the oldest leaves (Lyon *et al.*, 1943), 0.6–0.7 μg g^{-1} (Johnson *et al.*, 1952) and 0.3–0.4 μg g^{-1} (Stanton, 1966). However, Wallihan and Bradford (1977) found only 0.2 μg g^{-1} Mo in seedlings which appeared to be healthy. Leaves from deficient plants were found to contain 0.26 μg g^{-1} Mo (Lyon *et al.*, 1943) and 0.13 μg g^{-1} (Johnson *et al.*, 1952), though Wallihan and Bradford (1977) were unable to detect Mo in deficient seedlings. As the concentration in deficient leaves is very low, some of the differences between the values reported may reflect the difficulties of detecting small amounts of Mo with the analytical methods available.

References

Adams, P. (1978a). Tomatoes in peat. Part 1. How feed variations affect yield. *Grower* **89**, 1091, 1093–1094, 1097.

Adams, P. (1978b). Tomatoes in peat. Part 2. Effects of nutrition on tomato quality. *Grower* **89**,, 1142–1143, 1145.

Adams, P., Davies, J. N. and Winsor, G. W. (1978a). Effects of nitrogen, potassium and magnesium on the quality and chemical composition of tomatoes grown in peat. *J hort Sci* **53**, 115–122.

Adams, P., Graves, C. J. and Winsor, G. W. (1978b). Tomato yields in relation to the nitrogen, potassium and magnesium status of the plants and of the peat substrate. *Pl Soil* **49**, 137–148.

Adams, P., Graves, C. J. and Winsor, G. W. (1978c). Effects of copper deficiency and liming on the yield, quality and copper status of tomatoes, lettuce and cucumbers grown in peat. *Scientia Hort* **9**, 199–205.

Adams, P. and Winsor, G. W. (1974). Some responses of glasshouse tomatoes to boron. *J hort Sci* **49**, 355–363.

Adams, P., Winsor, G. W. and Donald, J. D. (1973). The effects of nitrogen, potassium and sub-irrigation on the yield, quality and composition of single-truss tomatoes. *J hort Sci* **48**, 123–133.

Adatia, M. H. and Winsor, G. W. (1971). Magnesium deficiency in glasshouse tomatoes. *Rep Glasshouse Crops Res Inst 1970*, 186–192.

Arnon, D. I. and Stout, P. R. (1939a). The essentiality of certain elements in minute quantity for plants with special reference to copper. *Pl Physiol* **14**, 371–375.

Arnon, D. I. and Stout, P. R. (1939b). Molybdenum as an essential element for higher plants. *Pl Physiol* **14**, 599–602.

Bailey, L. F. and McHargue, J. S. (1943). Copper deficiency in tomatoes. *Am J Bot* **30**, 558–563.

Bar-Akiva, A. (1964). Visible symptoms and chemical analysis vs. biochemical indicators as a means of diagnosing iron and manganese deficiencies in citrus plants. In Bould, C., Prevot, P. and Magness, J. R. eds. Plant Analysis and Fertiliser Problems IV, *Am Soc hort Sci*, East Lansing, 9–24.

Barke, R. E. and Menary, R. C (1971). Calcium nutrition of the tomato as influenced by total salts and ammonium nutrition. *Aust J exp Agric Anim Husb* **11**, 562–569.

Barker, A. V., Lackman, W. H., Maynard, D. N. and Puritch, G. S. (1967). Anatomical studies of ammonium-induced stem lesions in tomato. *HortSci* **2**, 159–160.

Bergmann, W. (1976). Ernährungsstörugen bei Kulturpflanzen in Farbbildern. *VEB Gustar Fisher Verlag, Jena*, 183pp.

Besford, R. T. (1975). Enzymes as indicators of crop nutritional status. *Rep Glasshouse Crops Res Inst 1974*, 59–60.

Besford, R. T. (1979a). Uptake and distribution of phosphorus in tomato plants. *Pl Soil* **51**, 331–340.

Besford, R. T. (1979b.). Effect of phosphorus nutrition in peat on tomato plant growth and fruit development. *Pl Soil* **51**, 341–353.

Besford, R. T. and Maw, G. A. (1975). Effect of potassium nutrition on tomato plant growth and fruit development. *Pl Soil* **42**, 395–412.

Bewley, W. F. and White, H. L. (1926). Some nutritional disorders of the tomato. *Ann Appl Biol* **13**, 323–338.

Borkowski, J. and Ostrzycka, J. (1973). The control of blossom-end rot of tomato and tipburn in lettuce by using the proper fertilisation. *Acta hort* **29**, 327–339.

Borkowski, J. and Szwonek, E. (1979). The influence of magnesium fertilisation on tomato fruit yield and magnesium nutritional status. *Proc 1st Int Symp Pl Nutr, Varnia (1979)* **1**, 372–382.

Bradfield, E. G. and Guttridge, C. G. (1984). Effects of night-time humidity and nutrient solution concentration on the calcium content of tomato fruit. *Scientia Hort* **22**, 207–217

Bradley, G. A. and Fleming, J. W. (1960). The effects of position of leaf and time of sampling on the relationship of leaf phosphorus and potassium to yield of cucumbers, tomatoes and watermelons. *Proc Am Soc hort Sci* **75**, 617–624.

Brennan, E. G. and Shive, J. W. (1948). Effect of calcium and boron nutrition of the tomato on the relation between these elements in the tissues. *Soil Sci* **66**, 65–75.

Brown, J. C. (1979). Effects of boron stress on copper enzyme activity in tomato. *J Pl Nutr* **1**, 39–53.

Bunt, A. C. (1969). Peat-sand substrates for plants grown in containers. I. The effect of base fertilisers. *Pl Soil* **31**, 97–110.

Bunt, A. C. (1976). Chapter 5, Nitrogen. In Modern Potting Compost. *George Allen and Unwin Ltd, Lond*, 82–106.

Burleson, C. A., Dacus, A. D. and Gerard, C. J. (1961). The effect of phosphorus fertilisation on the zinc nutrition of several irrigated crops. *Proc Soil Sci Soc Am* **25**, 365–368.

Cannell, G. H., Bingham, F. T. and Garber, M. J. (1960). Effects of irrigation and phosphorus on vegetative growth and nutrient composition of tomato leaves. *Soil Sci* **89**, 53–60.

Carolus, R. L., Erickson, A. E., Kidder, E. H. and Wheaton, R. Z. (1965). The interaction of climate and soil moisture on water use, growth and development of the tomato. *Q Bull Mich St Univ agric Exp Stn* **47**, 542–581.

Castro, P. R. C. and Malavolta, E. (1976). [The occurrence of blossom-end rot in tomatoes under the effect of growth regulators]. *Anais da Escola Superior de Agricultura "Luiz de Queiroz"* **33**, 173–189; *Hort Abstr* (1979) **49**, Abstr No 4271

Chapman, H. D. ed. (1966). Diagnostic Criterea for Plants and Soils. Berkeley *Univ Calif Div Agric Sci*, p713.

Chiu, T. and Bould, C. (1976). Effects of shortage of calcium and other cations on ^{45}Ca mobility, growth and nutritional disorders of tomato plants *(Lycopersicon esculentum)*. *J Sci Fd Agric* **27**, 969–977.

Christiansen, M. N. and Foy, C. D. (1979). Fate and function of calcium in tissue. *Commun Soil Sci Pl Anal* **10**, 427–442.

Clarke, E. J. (1944). Studies on tomato nutrition. I. The effect of varying concentrations of potassium on the growth and yields of tomato plants. *J Dep Agric Repub Ire* **41**, 53–81.

Clay, D. W. T. and Hudson, J. P. (1960). Effects of high levels of potassium and magnesium sulphates on tomatoes. *J hort Sci* **35**, 85–97.

Cornillon, P. (1974). Comportement de la tomate en fonction de la temperature du substrat. *Ann Agron* **25**, 753–777.

Cromwell, B. T. and Hunter, J. G (1942). Chlorosis in tomatoes. *Nature, Lond* **150**, 606–607.

Court, M. N., Stephen, R. C. and Waid, J. S. (1964a). Toxicity as a cause of the inefficiency of urea as a fertiliser. I. Review. *J Soil Sci* **15**, 42–48.

Court, M. N., Stephen, R. C. and Waid, J. S. (1964b). Toxicity as a cause of the inefficiency of urea as a fertiliser. II. Experimental. *J Soil Sci* **15**, 49–65.

Davies, J. N. (1957). Steam sterilisation studies. *Rep Glasshouse Crops Res Inst 1954/5*, 70–79.

Davies, J. N. (1964). Effect of nitrogen, phosphorus and potassium fertilisers on the non-volatile organic acids of tomato fruit. *J Sci Fd Agric* **15**, 665–673.

Davies, J. N. and Winsor, G. W (1967). Effect of nitrogen, phosphorus, potassium, magnesium and liming on the composition of tomato fruit. *J Sci Fd Agric* **18**, 459–466.

Dennis, D. J. (1968). Manganese toxicity in tomato seedlings. *NZ Comml Grow* **24** (6), 37, 39.

DeKock, P. C. (1983). Plant Physiology: Active Iron. *Rep Macaulay Inst Soil Res 1981/82* **52**, 84-85.

Drake, M. and White, J. M. (1961). Influence of nitrogen on uptake of calcium. *Soil Sci* **91**, 66-69.

Evans, H. J. and Troxler, R. V. (1953). Relation of calcium nutrition to the incidence of blossom-end rot in tomatoes. *Proc Am Soc hort Sci* **61**, 346-352.

Fisher, P. L. (1935). Responses of the tomato in solution cultures with deficiencies and excess of certain essential elements. *Md agric Exp Stn Bull no* **375**, 283-298.

Forster, W. A. (1951). Some effects of metals in excess on crop plants grown in soil culture. II. Effects of copper and zinc on crop plants grown in a variety of soils. *Rep Long Ashton Res Stn 1950*, 108-114.

Gerard, C. J. and Hipp, B. W. (1968). Blossom-end rot of 'Chico' and 'Chico grande' tomatoes. *Proc Am Soc hort Sci* **93**, 521-531.

Gomez-Lepe, B. E. and Ulrich, A. (1974). Influence of nitrate on tomato growth. *J Am Soc hort Sci* **99**, 45-49.

Goor, B. J., van, (1968). The role of calcium and cell permeability in the disease blossom-end rot of tomatoes. *Physiologia Pl* **21**, 1110-1121.

Graves, C. J., Adams, P., Winsor, G. W. and Adatia, M. H. (1978). Some effects of micronutrients and liming on the yield, quality and micronutrient status of tomatoes grown in peat. *Pl Soil* **50**, 343-354.

Gupta, U. C. (1983). Boron deficiency and toxicity symptoms for several crops as related to tissue boron levels. *J Pl Nutr* **6**, 387-395.

Hall, D. A. (1977). Some effects of varied calcium nutrition on the growth and composition of tomato plants. *Pl Soil* **48**, 199-211.

Hamson, A. R. (1952). Factors which condition firmness in tomatoes. *Food Res* **17**, 370-379.

Hester, J. B. (1938). A trace element deficiency on the tomato. *Proc Am Soc hort Sci* **36**, 744-746.

Hewitt, E. J. (1944). Experiments in mineral nutrition. I. The visual symptoms of mineral deficiencies in vegetables and cereals grown in sand cultures. *Rep Long Ashton Res Stn 1943*, 33-47.

Hewitt, E. J. (1945). Experiments in mineral nutrition. II. The visual symptoms of mineral deficiencies in crop plants grown in sand cultures. *Rep Long Ashton Res Stn 1944*, 50-60.

Hewitt, E. J. and Bolle-Jones, E. W. (1952). Molybdenum as a plant nutrient. II. The effects of molybdenum deficiency on some horticultural and agricultural crop plants in sand culture. *J hort Sci* **27**, 257-265.

Hewitt, E. J., Bolle-Jones, E. W. and Miles, P. (1954). The production of copper, zinc and molybdenum deficiences in crop plants grown in sand culture with special reference to some effects of water supply and seed reserves. *Pl Soil* **5**, 205-222.

Hewitt, E. J. and Watson, E. F. (1980). The production of microelement deficiencies in plants grown in recirculated nutrient film (NFT) systems. *Acta Hort* **98**, 179-189.

Hobson, G. E., Davies, J. N. and Winsor, G. W. (1977). Ripening disorders of tomato fruit. Growers' Bull no 4, Glasshouse Crops Res Inst, Littlehampton, 7-8.

Hogue, E., Wilcox, G. E. and Cantliffe, D. J. (1970). Effect of soil phosphorus levels on phosphate fractions in tomato leaves. J Am Soc hort Sci 95, 174-176.

Humphries, P. D. and Devonald, V. G. (1977). The distribution of potassium, calcium and magnesium in young tomato plants grown in water culture. Pl Soil 48, 435-445.

Johannesson, J. K. (1951). Magnesium deficiency in tomato leaves. NZ J Sci Tech, Ser A 33, 52-57.

Johnson, C. M., Pearson, G. A. and Stout, P. R. (1952). Molybdenum nutrition of crop plants. II. Plant and soil factors concerned with molybdenum deficiencies in crop plants. Pl Soil 4, 178-196.

Johnston, E. S. and Dore, W. H. (1928). Relation of boron to the growth of the tomato plant. Science 67, 324-325.

Johnston, E. S. and Fisher, P. L. (1930. The essential nature of boron to the growth and fruiting of the tomato. Pl Physiol 5, 387-392.

Johnston, E. S. and Hoagland, D. R. (1929). Minimum potassium level required by tomato plants grown in water cultures. Soil Sci 27, 89-109.

Jones, J. O., Nicholas, D. J. D., Wallace, T. and Jefferiss, A. (1945). Experiments on the control of magnesium deficiency in glasshouse tomatoes. Progress Report 11. Rep Long Ashton Res Stn 1944, 61-71.

Kertesz, Z. I. (1939). Effect of added calcium on canned tomatoes. Canner 88, No. 24, 14-16.

Kirkby, E. A. and Mengel, K. (1967). Ionic balance in different tissues of the tomato plant in relation to nitrate, urea or ammonium nutrition. Pl Physiol 42, 6-14.

Kitchen, H. B. (1948). Diagnostic Techniques for Soils and Crops. The American Potash Institute Washington DC, 308pp.

Kloes, L. J. J., van der, Boon, J., van der, Delver, P., Knoppien, P., Pouwer, A., Roorda van Eysinga, J. P. N. L. and Schouwenburg, J. C., van, (1961). Eenjarige bemestingsproeven met stikstof bij koud geteelde tomaten. Institute Bodemvruchtbaarheid, Groningen, Rapport 6.

Lamb, J. G. D. (1961). A case of manganese toxicity affecting a cold house tomato crop. Ir J agric Res 1, 17-20.

Lamb, J. G. D. and Conroy, E. (1962). Minor element deficiencies noted in commercial tomato crops in Ireland during 1961. Ir J agric Res 1, 342-343.

Lingle, J. C., Holmberg, D. M. and Zobel, M. P. (1958). The correction of zinc deficiency of tomatoes in California. Proc Am Soc hort Sci 72, 397-402.

Locascio, S. J. and Warren, G. F. (1960). Interaction of soil temperature and phosphorus on growth of tomatoes. Proc Am Soc hort Sci 75, 601-610.

Lyon, C. B., Beeson, K. C. and Ellis, G. H. (1943). Effects of micro-nutrient deficiencies on growth and vitamin content of the tomato. Bot Gaz 104, 495-514.

MacGillivray, J. H. (1927). Effect of phosphorus on the composition of the tomato plant. J agric Res 34, 97-127.

MacKay, D. C., Langille, W. M. and Chipman, E. W. (1962). Boron deficiency and toxicity in crops grown on sphagnum peat soil. Can J Soil Sci 42, 302-310.

MacLean, K. S., McLaughlin, H. A. L. and Brown, M. H. (1968). The application of tissue analysis to the production of commercial greenhouse tomatoes. Proc Am Soc hort Sci 92, 531-536.

Maher, M. J. (1976). Growth and nutrient content of a glasshouse tomato crop grown in peat. Scientia Hort 4, 23-26.

Maher, M. J. (1978). The effect of root zone warming on tomatoes grown in nutrient solution at two air temperatures. Acta Hort 82, 113-117.

Majewski, F. and Majewska, W. (1953). Studies on the effect of boron on tomatoes. Rocznik Nauk Rolniczych, Series A, 68, 65-84; Hort Abstr 24, 398.

Malcolm, J. L. (1959). Effect of nitrogen, phosphorus and potassium fertiliser on fruit yield and composition of tomato leaves. J Agric Fd Chem 7, 415-418.

Markus, D. K. and Steckel, J. E. (1980). Periodical analysis of artificial rooting media and tomato leaf analysis from New Jersey greenhouses. Acta Hort 99, 205-217.

Martens, N. G. C. (1963). Experiments on the calcium nutrition of tomatoes. Jversl Roermond 1963, 19-20.

Martin, G. C. and Wilcox, G. E. (1963). Critical soil temperature for tomato plant growth. Proc Soil Sci Soc Am 27, 565-567.

Mason, S. C. and Wilcox, G. E. (1982). Nitrogen status evaluation of tomato plants. J Am Soc hort Sci 107, 483-486.

Massey, D. M. and Winsor, G. W. (1980a). Some responses of tomato to nitrogen in recirculating solutions. Acta Hort 98, 127-137.

Massey, D. M. and Winsor, G. W. (1980b). Some responses of tomato plants to phosphorus concentration in nutrient film culture. Proc 5th Int Congr Soilless Culture Wageningen 1980, 205-214. Wageningen ISOSC.

Maynard, D. N., Barker, A. V. and Lachman, W. H. (1968). Influence of potassium on the utilisation of ammonium by tomato plants. Proc Am Soc hort Sci 92, 537-542.

Maynard, D. N., Gersten, B. and Michelson, L. F. (1959). The effects of boron nutrition on the occurrence of certain tomato fruit disorders. Proc Am Soc hort Sci 74, 500-505.

Menary, R. C. and Kruger, N. S. (1966). Influence of soil bulk density on nutrition and growth in the tomato. Qd J agric Sci 23, 359-371.

Messing, J. H. L. (1957). Boron in the nutrition of tomatoes. Rep Glasshouse Crops Res Inst 1954/55, 79-84.

Messing, J. H. L. and Winsor, G. W. (1957). Boron in the nutrition of tomato plants. Rep Glasshouse Crops Res Inst 1956, 77-83.

Millikan, C. R. (1951). Radio-autographs of manganese in plants. Aust J sci Res, Ser B, biol Sci 4, 28-40.

Nicholas, D. J. D. (1951). Some effects of metals in excess on crop plants grown in soil culture. I. Effects of copper, zinc, lead, cobalt, nickel and manganese on tomato grown in an acid soil. Rep Long Ashton Res Stn 1950, 96-108.

Nightingale, G. T., Schermerhorn, L. G. and Robbins, W. R. (1932). Effects of sulphur deficiency on metabolism in tomato. Pl Physiol 7, 565-595.

Oertli, J. J. and Kohl, H. C. (1961). Some considerations about the tolerance of various plant species to excessive supplies of boron. Soil Sci 92, 243-247.

Pill, W. G., Lambeth, V. N. and Hinckley, T. M. (1978). Effects of nitrogen form and level on ion concentrations, water stress and blossom-end rot incidence in tomato. J Am Soc hort Sci 103, 265-268.

Puritch, G. S. and Barker, A. V. (1967). Structure and function of tomato leaf chloroplasts during ammonium toxicity. Pl Physiol 42, 1229-1238.

Purvis, E. R. and Carolus, R. L. (1964). Nutrient deficiencies in vegetable crops. In H.B. Sprague, ed. Hunger Signs in Crops, 3rd ed 257. David McKay Co N Y.

Raleigh, S. M. and Chucka, J. A. (1944). Effect of nutrient ratio and concentration on growth and composition of tomato plants and on the occurrence of blossom-end rot of the fruit. Pl Physiol 19, 671-678.

Reeve, E. and Shive, J. W. (1944). Potassium-boron and calcium-boron relationships in plant nutrition. Soil Sci 57, 1-14.

Riekels, J. W. and Gulen, O. F. (1967). Iron uptake and translocation by tomato plants as influenced by manganese. Pl Physiol 42, Suppl 6.

Roorda van Eysinga, J. P. N. L. (1966). Bemesting van tomaten met kali. *Versl landbouwk Onder* **667**, 37pp.

Roorda van Eysinga, J. P. N. L. (1971a). Fertilisation of tomatoes with nitrogen. *Agric Res Rep* 754, 17pp. Cent agric Publ Documn Wageningen.

Roorda van Eysinga, J. P. N. L. (1971b). Fertilization of tomatoes with phosphate. *Agric Res Rep* 755 10pp. Cent agric Publ Documn, Wageningen.

Roorda van Eysinga, J. P. N. L. and Smilde, K. W. (1981). Nutritional disorders in glasshouse tomatoes, cucumbers and lettuce. *Centre agric Publ Docum Wageningen*. 130pp.

Sanders, J. R. and Bloomfield, C. (1980). The influence of pH, ionic strength and reactant concentrations on copper complexing by humified organic matter. *J Soil Sci* **31**, 53-63.

Seaton, H. L. and Gray, G. F. (1936). Histological studies of tissues from greenhouse tomatoes affected by blotchy ripening. *J agric Res* **52**, 217-224.

Shafshak, S. A. and Winsor, G. W. (1964). A new instrument for measuring the compressibility of tomatoes, and its application to the study of factors affecting fruit firmness. *J hort Sci* **39**, 284-297.

Shaykewich, C. F., Yamaguchi, M. and Campbell, J. D. (1971). Nutrition and blossom-end rot of tomatoes as influenced by soil water regime. *Can J Pl Sci* **51**, 505-511.

Sheldrake, R. (1981). Money bags? *Am Veg Grow 29* (11), 15, 16, 34, 36.

Smilde, K. W. (1972). Trace nutrient requirements of some plant species on peat substrates. *Proc 4th Int Peat Congr Helsinki* **3**, 239-257.

Smilde, K. W. and Roorda van Eysinga, J. P. N. L. (1968). Nutritional diseases in glasshouse tomatoes. *Cent agric Publ Docum Wageningen*, 48pp.

Sommer, A. L. (1931). Copper as an essential for plant growth. *Pl Physiol* **6**, 339-345.

Sonneveld, C. and Voogt, S. J. (1980). The application of manganese in nutrient solutions for tomatoes grown in a recirculating system. *Acta Hort* **98**, 171-178.

Stanton, D. J. (1966). Report on tomato trials. *Cawthron Inst Trienn Rep 1963-66*, 25-27.

Starck, J. R., Grudzinska-Kuhn, J. and Wojciechowski, J. (1973). Leaf analysis as a means of determining the nutrient requirements of greenhouse tomatoes. *Acta Hort* **29**, 81-88.

Stromme, E. (1957). Avling og kvalitet av veksthustomater i ct faktorielt gjødslingsforsøk med N, K og Mg. *Forsk Fors Landbr* **8**, 447-446.

Swiader, J. M. and Morse, R. D. (1982). Phosphorus solution concentrations for production of tomato, pepper and eggplant in mine soils. *J Am Soc hort Sci* **107**, 1149-1153.

Torres de Claassen, M. E. and Wilcox, G. E. (1974). Effect of nitrogen form on growth and composition of tomato and pea tissue. *J Am Soc hort Sci* **99**, 171-174.

Twyman, E. S. (1951). The iron and manganese requirements of plants. *New Phytol* **50**, 210-226.

Uljee, A. H. (1964). Ammonium nitrogen accumulation and root injury to tomato plants. *NZ J agric Res* **7**, 343-356.

Wall, M. E. (1940). The role of potassium in plants: II. Effect of varying amounts of potassium on the growth status and metabolism of tomato plants. *Soil Sci* **49**, 315-331.

Wallace, T. (1951). The Diagnosis of Mineral Deficiencies in Plants by Visual Symptoms. 2nd ed. 107pp. Lond HMSO.

Wallihan, E. F. and Bradford, G. R. (1977). Simplified methods for inducing micronutrient deficiencies. *HortSci* **12**, 327-328.

Wallihan, E. F., Sharpless, R. G. and Printy, W. L. (1978). Cumulative toxic effects of boron, lithium and sodium in water used for hydroponic production of tomatoes. *J Am Soc hort Sci* **103**, 14-16.

Walsh, T. and Clarke, E. J. (1942). A chlorosis of tomatoes. *J Dep Agric, Eire* **39**, 316-325.

Walsh, T. and Clarke, E. J. (1945a). A chlorosis of tomatoes in relation to potassium and magnesium nutrition. *J R hort Soc* **70**, 202-207.

Walsh, T. and Clarke E. J. (1945b). Iron deficiency in tomato plants grown in an acid peat medium. *Proc R Ir Acad* **50**, Section B, No. 22, 359-372.

Ward, G. M. (1963). The application of tissue analysis to greenhouse tomato nutrition. *Proc Am Soc hort Sci* **83**, 695-699.

Ward, G. M. (1964). Greenhouse tomato nutrition – a growth analysis study. *Pl Soil* **21**, 125-133.

Ward, G. M. (1973). Causes of blossom-end rot of tomatoes based on tissue analysis. *Can J Pl Sci* **53**, 169-174.

Ward, G. M. (1976). Sulphur deficiency and toxicity symptoms in greenhouse tomatoes and cucumbers. *Can J Pl Sci* **56**, 133-137.

Ward, G. M. (1977). Manganese deficiency and toxicity in greenhouse tomatoes. *Can J Pl Sci* **57**, 107-115.

White, H. L. (1938). Observations of the effect of nitrogen and potassium on the fruiting of the tomato. *Ann Appl Biol* **25**, 20-49.

White, H. L. (1938). Further observations on the incidence of blotchy ripening of the tomato. *Ann Appl Biol* **25**, 544-557.

Widders, I. E. and Lorenz, O. A. (1982). Potassium nutrition during tomato plant development. *J Am Soc hort Sci* **107**, 960-964.

Wiersum, L. K. (1966). The calcium supply of fruits and storage tissues in relation to water transport. *Acta Hort* **4**, 33-38.

Wilcox, G. E., Hoff, J.E. and Jones, C. M. (1973). Ammonium reduction of calcium and magnesium content of tomato and sweet corn leaf tissue and influence on incidence of blossom-end rot of tomato fruit. *J Am Soc hort Sci* **98**, 86-89.

Winsor, G. W. (1966). A note on the rapid assessment of "boxiness" in studies of tomato fruit quality. *Rep Glasshouse Crops Res Inst 1965*, 124-127.

Winsor, G. W. (1970). A long-term factorial study of the nutrition of greenhouse tomatoes. *Proc 6th Colloq Int Pot Inst, Florence, 1968*, 269-281.

Winsor, G. W. (1969). Plant nutrition: the special needs of glasshouse crops. *Span* **12**, 154-158.

Winsor, G. W. (1973). Nutrition. In: The UK tomato manual. *Grower Books Lond*, 35-42.

Winsor, G. W. (1979). Some factors affecting the quality and composition of tomatoes. *Acta Hort* **93**, 335-346.

Winsor, G. W. and Adams, P. (1968). The nutrition of glasshouse crops. *Rep Glasshouse Crops Res Inst 1967*, 65-70.

Winsor, G. W., Davies, J. N. and Long, M. I. E. (1961). Liquid feeding of glasshouse tomatoes: the effects of potassium concentration on fruit quality and yield. *J hort Sci* **36**, 254-267.

Winsor, G. W., Davies, J. N. and Long, M. I. E. (1967). The effects of nitrogen, phosphorus, potassium, magnesium and lime in factorial combination on the yields of glasshouse tomatoes. *J hort Sci* **42**, 277-288.

Winsor, G. W., Davies, J. N. and Messing, J. H. L. (1958). Studies on potash/nitrogen ratio in nutrient solutions, using trickle irrigation equipment. *Rep Glasshouse Crops Res Inst 1957*, 91-98.

Winsor, G. W., Davies, J. N., Messing, J. H. L. and Long, M. I. E. (1962). Liquid feeding of glasshouse tomatoes; the effects of nutrient concentration on fruit quality and yield. *J hort Sci* **37**, 44-57.

Winsor, G. W., Hurd, R. G. and Price, D. (1979). Nutrient film technique. Growers' Bull No 5. *Glasshouse Crops Res Inst Littlehamptom W Sussex*, 48pp.

Winsor, G. W. and Long, M. I. E. (1963). Some effects of potassium and lime on the relation between phosphorus in soil and plant, with particular reference to glasshouse tomatoes, carnations and winter lettuce. *J Sci Fd Agric* **14**, 251-259.

Winsor, G. W. and Long, M. I. E. (1967). The effects of nitrogen, phosphorus, potassium, magnesium and lime in factorial combination on ripening disorders of glasshouse tomatoes. *J hort Sci* **42**, 391-402.

Winsor, G. W. and Long, M. I. E. (1968). The effects of nitrogen, phosphorus, potassium, magnesium and lime in factorial combination on the size and shape of glasshouse tomatoes. *J hort Sci* **43**, 323-334.

Winsor, G. W. and Massey, D. M. (1978). Some aspects of the nutrition of tomatoes grown in recirculating solution. *Acta Hort* **82**, 121-132.

Winsor, G. W., Messing, J. H. L., Hobson, G. E. and Long, M. I. E. (1965b). The magnesium content and potassium magnesium ratio of tomato leaves in relation to degree of chlorosis. *J hort Sci* **40**, 156-166.

Winsor, G. W., Messing, J. H. L. and Long, M. I. E. (1965a). The effects of magnesium deficiency on the yield and quality of glasshouse tomatoes grown at two levels of potassium. *J hort Sci* **40**, 118-132.

Woolley, J. T. and Broyer, T. C. (1957). Foliar symptoms of deficiencies of inorganic elements in tomato. *Pl Physiol* **32**, 148-151.

CHAPTER 6
Carnation

On new nurseries, carnations are sometimes planted directly in the glasshouse soil. The plant roots may reach a depth of more than 60 cm (Ministry of Agriculture, 1967), and are thus not subject to the wide fluctuations of moisture content or nutrient levels which can occur in shallower beds.

Carnations are, however, very susceptible to soil-borne diseases such as the wilt caused by *Fusarium oxysporum f. dianthi*. Since carnation crops normally occupy the beds for periods of 1 – 2 years and are grown in monoculture, the spread of such diseases can prove disastrous. Thus Williamson (see Langhans, 1961, p. 60) noted that, unless soil-borne diseases are controlled, '. . . plant growing becomes a haphazard operation filled with risk and uncertainty'. Soil sterilisation cannot be relied upon to provide more than temporary control, since re-infestation from below is almost inevitable. The most satisfactory alternative is to grow the crop in beds isolated from the underlying soil by polythene sheet, asbestos sheet or concrete. The shallow nature of such beds, usually 15–20 cm, gives a relatively restricted root zone. Special attention is therefore required to maintain desirable levels of moisture and nutrients at all times, for which purposes trickle or low level irrigation systems are particularly appropriate. The soil used in separated beds is usually amended with peat and grit to improve aeration and water retention. Other materials such as perlite (a heat-expanded volcanic mineral) are sometimes incorporated, and soilless substrates such as peat/sand mixtures may be used instead of soil.

The carnation is often regarded as a rather tolerant crop with regard to nutrition. The nature of the foliage is such that chlorosis or variations in leaf colour due to mineral deficiencies or excesses are less obvious than in many other glasshouse crops. The long growth period of the crop has doubtless limited the number of major nutritional studies undertaken with it. Furthermore, being propagated vegetatively and relatively slow growing, the carnation has not commended itself to research workers merely needing a convenient 'test plant'. Published data on the nutrition of greenhouse carnations are thus somewhat limited. Long-term crop responses to nutrient application have, however, been studied factorially in the UK (e.g. Winsor *et al.*, 1970; Adams *et al.*, 1979) and much useful information is available in the bulletins of the various American growers' associations. Leaf analysis as a guide to the nutritional requirements of carnations has

attracted particular attention in the USA, as may be seen from the analytical references cited in Tables 22 and 23. These two Tables contain a selection of recommendations and experimental ranges for groups of three or more elements; further data are referred to in the text. A detailed study of the distribution of N, P, K, Ca, Mg and B from the base to the tip of carnation leaves, and also of the effect of sampling position on the shoot, was made by Nelson and Boodley (1963). Background information on carnation growing, including general crop nutrition, will be found in manuals by Langhans (1961), Holley and Baker (1963) and the (UK) Ministry of Agriculture (1967). Recommended rates of application of base fertilisers for carnations in relation to pre-planting soil analyses were published by the Ministry of Agriculture (1982). Aspects of the nutrition of commercial carnation crops requiring particular attention include (a) maintenance of an adequate level of B to decrease calyx splitting, particularly in calcareous soils, and (b) addition of supplementary dressings of P during the second season of cropping.

Nitrogen

Symptoms and growth responses. The visual symptoms of N deficiency have been described in detail by Messing and Owen (1952), Messing (1953, 1955, 1958), based on studies in sand culture. Growth was decreased by omission and few axillary shoots were produced. The internodes failed to elongate normally, and the older leaves turned yellow. Necrosis started at the tips of the lower leaves and progressed towards the leaf base across the whole width of the lamina; the dead tissue was of a pale straw colour. Holley (1956) reported that, at low N levels, the leaves became thin and narrow and lost their curl; development of lateral shoots was retarded and the colour became dull or yellow-green.

Symptoms described as 'curly-tip' were recorded by Messing (1953, 1958); the tips of the young leaves failed to separate on N deficient plants, and continuing growth caused the characteristic curvature illustrated by Messing (1955, 1958). Carnation plants deprived of N at various stages of growth and times of the year invariably developed 'curly tip', but the effect was most pronounced under conditions of low (winter) light. This association of 'curly tip' with N deficiency during winter was confirmed by Winsor and Long (1962). Messing (1953) also recorded rather similar symptoms in plants deprived of S, though

Source		Nutrient content of dry matter (%)				
		N	P	K	Ca	Mg
White (1966)[a]	Mean	3.71	0.36	3.09	1.38	0.42
	Range	3.4–3.9	0.34–0.41	2.8–3.3	1.3–1.5	0.19–0.61
Holley (1968); see also Hanan (1975)		3.2–3.6	0.20–0.35	2.9–3.3	1.5–2.0	0.20–0.40
Boodley, quoted by Holley (1968)	–	3.0–5.0	0.20–0.30	2.0–6.0	1.0–2.0	0.20–0.35
Winsor et al. (1970)	Deficiency	2.0–2.3	0.1–0.15	1.5–2.0	–	–
	Favourable	3.0–3.5	0.25–0.30	3.0–4.0	–	–
Criley and Carlson (1970)[b]	Optimal	4.2	0.25	4.2	1.5	0.38
Puustjärvi (1972)	–	3.40	0.26	4.40	0.43	0.60
Khattab et al. (1977)	Vegetative	2.5–3.7	0.2–0.7	2.4–4.2	–	–
	Flowering	2.52–3.88	0.35–0.44	4.0–6.0	–	–
Kazimirova (1977)	1st year	2.0–2.7	0.3–0.5	2.9–6.3	–	–
	2nd year	2.0–2.2	0.2–0.3	2.8–3.6	–	–
Fortney and Wolf (1981)	Normal	3.2–5.2	0.20–0.35	2.5–6.0	1.0–2.0	0.25–0.50
	Deficiency	<3.0	<0.05	<2.0	<0.60	<0.15
Puustjärvi (1981b)	–	3.5–5.0	0.4–0.5	3.0–5.0	1.4–1.9	0.38–0.46
Peterson (1982)[c]	Excess	5.26	0.36	6.11	2.11	0.56

Table 22. Macronutrient content of carnation leaves.

[a] Based on 360 samples from 5 growers.
[b] Values quoted by Criley and Carlson from a commercial source.
[c] Peterson's interpretative ranges for sufficiency and deficiency are very similar to those quoted by Fortney and Wolf, and only the threshold values for excess levels have therefore been included in the Table.

Source		Nutrient content of dry matter ($\mu g\ g^{-1}$)				
		Fe	Mn	Cu	Zn	B
White (1966)[a]	Mean	73	93	7	37	37
	Range	54–87	52–186	5–11	24–48	28–53
Boodley, quoted by Holley (1968)		50–150	100–300	10–30	25–75	25–400
Criley and Carlson (1970)[b]	Optimal	100	200	20	50	65
Parker (1971a)	Low	30–50	20–50	4–5	18–25	–
	Adequate	50–100	50–150	5–10	25–100	–
	High	150–200	1200–1500	10–20	150–300	–
Puustjärvi (1971)[c]		80	420	20	–	–
Hanan (1975)[d]		50–100	50–150	5–10	25–100	25–100
Fortney and Wolf (1981)	Normal	50–150	100–300	10–30	25–75	30–100
	Deficient	<30	<30	<5	<15	<25
Puustjärvi (1981a)	Optimum	50–200	30–250	5–15	15–50	30–60
Peterson (1982)[c]	Excess	156	800	36	81	700

Table 23. Micronutrient content of carnation leaves.

[a] Based on 360 samples from 5 commercial growers.
[b] Quoted by Criley and Carlson from a commercial source.
[c] Average value for experimental plants grown in peat.
[d] Recommended values.
[e] Interpretative ranges for sufficiency and deficiency have been omitted since these were similar to those published by Fortney and Wolf (1981).

this deficiency has not been reported for commercial crops. 'Curly tip' symptoms arise from failure of the inward-rolled leaf tips to flatten and separate, however, and S deficiency is known to cause stiffness in the leaves (Messing, 1953).

The effects of N concentration (25, 125 and 225 mg l^{-1} N) in the irrigation water were studied by Winsor and Long (1962) in a factorial experiment with two carnation cultivars (Improved William Sim and Saugus White) grown in soil with added peat and grit. The total number of blooms produced increased by 6–9% at the intermediate and high levels compared with the lowest level of this nutrient (P < 0.001). Splitting of the calyx was decreased markedly by high N levels, the percentages for cv. 'Saugus White' ranging from 15.4% at 25 mg l^{-1} N to 6.9% at 225 mg l^{-1} N. A second variety, 'Improved William Sim', produced a very high proportion of split calcyces in this trial (34.6% overall), but still showed a significant reduction in splitting from 45.1% at 25 mg l^{-1} N to 27.3% at 225 mg l^{-1}. These results support the earlier work of Beach (1952) but not that of Clapp and Folley (1941), who found 43% split calyces at high N compared with 18% at low N. The responses of two Sim varieties to N concentration (20, 90, 160 and 230 mg l^{-1} N) in the irrigation water were reported by Winsor et al. (1970). Flower production increased with N concentration up to 160 mg l^{-1} N in the first experiment and up to 90 mg l^{-1} N in the second. In both experiments, each covering a period of 22 months, the proportion of blooms with split calyces decreased progressively with increasing N concentration (Table 24).

The effects of a more restricted range of concentrations in the irrigation water (100, 140, 180, 220 and 260 mg l^{-1} N) on the flowering of carnations were later studied by Adams et al. (1979). The treatments were tested on two 'Sim' cultivars, grown at two spacings (10 cm × 20 cm and 20 cm × 20 cm) over a period of 22 months. The numbers of blooms produced were not affected significantly by concentration within this range, confirming that most of the flowering response to N occurs below 100 mg l^{-1} N. The quality of blooms was, however, markedly affected by concentration. Thus splitting of the calyx decreased significantly (P < 0.001) with increasing N level, the mean values ranging from 14.0% at 100 mg l^{-1} N to 8.7% at 260 mg l^{-1} N. A significant interaction between B and N was also found; the effect of N on splitting occurred mainly in the absence of added B, under which conditions splitting decreased from 19.1% at 100 mg l^{-1} N to 9.9% at 260 mg l^{-1} N.

Although the effects of N concentration on splitting (Adams, et al., 1979) could partly be explained by an indirect effect on soil pH and hence on the availability of B, significant responses were reported by Winsor et al., (1970) even when B was included in all the irrigation water throughout the trial. Regression analysis (Adams et al., 1979) showed that addition of a linear term for N concentration in the nutrient solutions to the quadratic

| Experiment | Cultivar | Nitrogen concentration (mg l^{-1} N) | | | | LSD* (P=0.05) |
		20	90	160	230	
A	William Sim	19.2	12.9	9.5	6.8	2.4
	Pink Sim	23.6	15.5	14.0	11.0	
	Mean	21.4	14.2	11.8	8.9	1.9
B	William Sim	12.9	7.9	6.2	4.7	2.6
	Pink Sim	17.5	10.4	9.3	7.7	
	Mean	15.2	9.2	7.8	6.2	2.5

Table 24. Effect of nitrogen on the percentage of blooms with split calyces (Winsor et al., 1970).
* For comparison of N levels.

regression of calyx splitting (%) on soil pH increased the proportion of variance accounted for from 40% to 70%. High N levels were thus associated with a reduction in splitting even when the effects of soil pH had been taken into account.

Several workers have discussed the effects of N supplied to carnations as NH_4–N and NO_3–N. According to Hartman and Holley (1968), NH_4–N should be kept low for growth in inert media whereas up to half of the N supplied to soil-grown crops may be in this form. Schekel (1971) found that both flower production and vegetative growth (fresh weight) increased with NH_4–N concentration up to 4 meq l^{-1} N (56 mg l^{-1}), this being the highest level tested. The plants (cv. White Sim) were grown in gravel at low light intensities (November to April). It was suggested that, under these conditions, nitrate reductase activity might have been low, but the author found little accumulation of NO_3–N in the tissue to support this hypothesis. Green and Holley (1974) concluded that, during periods of low solar radiation, the optimum ratio of NH_4–N to NO_3–N was 1:2, whereas an 'all-nitrate' feed was preferred at high radiation levels. It was suggested that NO_3–N reduction served as an alternative sink for excess photosynthetic energy.

Carnations grown in soil sterilised at 100°C and fed with NH_4–N showed loss of roots, interveinal chlorosis and wilting associated with high concentrations of soluble and exchangeable Mn and NH_4-N (White, 1968).

Nitrogen content of the leaves. Some published data on the N content of carnation leaves are summarised in Table 22. Most authors have suggested values between 3.0 and 3.9% N as favourable to the crop. Lower values were, however, included in the ranges reported by Khattab et al. (1977), namely 2.5–3.7% N for the leaves of vegetative shoots and 2.5–3.9% N for flowering shoots. Still lower values were recorded by Kazimirova (1977), who found 2.0–2.7% N and 2.0–2.2% N in the leaves during the first and second years of growth. Such values are similar to those regarded as deficient by Winsor et al. (1970), who observed symptoms below 2.0–2.3% N in the leaves. The values reported by Fortney and Wolf (1981) were a little higher than most, the 'normal range' extending up to 5.2% N and deficiency being indicated below 3.0%. Puustjärvi (1981b) recommended 3.5–5.0% N in the leaves.

Nelson and Boodley (1965a,b) reported a complex study of plant response to N concentration in solution over the range 5–900 mg l⁻¹ N. The nutrients were applied at every watering to carnations ('Improved White Sim') grown in a mixture of loam, peat, perlite and sand (9:6:4:2). Quality index values were obtained by averaging the scores obtained from a multi-component scheme of visual grading and by standard commercial grading; the corresponding analytical values ('standard tissue concentrations') were plotted against time of year. For a June-planted crop the preferred N content rose from about 3.1% in August to 5.0% in March, falling sharply to about 3.5% in mid-summer and then rising steadily to 4.3% by the following spring. Some of these values, interpolated from the published graph, are relatively high compared with those of other workers.

Parker and Holley (1972) concluded that determination of NO_3–N in the leaves did not give a satisfactory indication of N deficiency. Nelson and Boodley (1965b) reported that the Littlefield and miniature varieties had lower N contents in their leaves than the more commonly grown 'Sim' varieties.

Phosphorus

Symptoms and growth responses. Phosphorus deficiency greatly reduced the growth of carnations, giving narrow leaves and small flowers (Holley, 1956). Messing and Owen (1952) found no characteristic symptoms by which the deficiency could readily be identified; the plants were stunted and the flowers were slightly smaller than where a complete nutrient solution was supplied. They recorded, however, that the old leaves' . . . died at an abnormal rate and in an irregular manner, most often starting at the base of the leaf and eventually becoming brown'. Plants deprived of P at an early stage in sand culture had short internodes and produced few sideshoots (Messing, 1953). Irregular chlorotic patches developed on any part of the leaf, followed by a spreading necrosis; similar symptoms were illustrated by Penningsfeld (1970).

Winsor et al. (1970) reported that, where P was omitted from the fertiliser treatments for soil-grown carnations, the growth of two Sim varieties was restricted and 'spiky'. The leaves were narrow and silvery blue-green colour. The basal leaves died off progressively, often from the tips, and in extreme instances only the tops of the plants remained green during the second season of cropping. The position of the symptoms doubtless arises from mobilisation of P in the older leaves and translocation towards the actively-growing tops of the plants.

Incorporation of superphosphate (7.9% P) in a soil-based substrate at the rate of 5.35 kg m⁻³ increased flower production by cv. Improved William Sim by 20% (P < 0.001) but was not beneficial to cv. Saugus White (Winsor and Long, 1962). Further examination of the data showed that superphosphate had reduced flowering of this latter variety in the unlimed plots, though not where lime had been incorporated. This varietal interaction was doubtless associated with the acidity of superphosphate in a steamed, soil-based compost; tip-burn of the leaves developed on cv. Saugus White where superphosphate had been used, but the Sim variety was unaffected.

Two Sim varieties (William Sim and Pink Sim), grown in two successive experiments at four levels of P, showed progressive increases in flower production over the range 0–3.0 kg superphosphate (7.9% P) per cubic metre of substrate (Winsor et al., 1970). The overall responses in flower production were 27% and 67% respectively in the first and second crops, with still greater gains (38% and 88%) when plots deficient in N and K were excluded.

Phosphorus content of the leaves. As shown in Table 22, numerous authors have published favourable or typical values of P in carnation leaves. Many of the values fall within the range 0.2–0.35% P. Where data were reported for both the first and second seasons of crop growth (Winsor et al., 1970; Kazimirova, 1977), the higher values were found in the first year. Still higher levels have sometimes been reported for carnation leaves, with values up to 0.7% in young vegetative shoots (Khattab et al., 1977). Hanan (1975) found 0.45–0.65% P in experimental plants grown in gravel. Particularly high P levels were recorded for very young plants by Parker and Holley (1972): thus, eight weeks after planting, the 'standard' plants receiving complete nutrient solutions contained 1.01% P in the leaves, decreasing to 0.60% and 0.45% after 12 and 16 weeks. Even higher values (1.75% P) were found in young plants grown for 8 weeks at low Mg (1 meq l⁻¹), though the content decreased rapidly thereafter, reaching 0.47% P in the 16th week of treatment.

Winsor et al. (1970) reported obvious symptoms of deficiency at values below 0.1–0.15% P in the leaves. Fortney and Wolf (1981) quoted a far lower value (0.05% P) below which plants would probably exhibit deficiency symptoms resulting in reduction of growth and quality.

POTASSIUM

Symptoms of deficiency. The symptoms of K deficiency were described by Szendel (1940). The upper and middle leaves developed yellowish necrotic spots, particularly at the tip and along the upper margins; the affected region became dried and shrivelled. Clapp and Folley (1941) found 'no clearly defined symptoms' of K deficiency where K was omitted in sand culture, but recorded slender and shortened stems, narrow leaves and small blooms. Further studies were reported by Messing and Owen (1952) and Messing (1953, 1958). The tips of leaves on plants grown without K showed slight loss of colour followed (particularly in Sim varieties) by rapid development of light-coloured sunken spots of irregular shape and size. With cv. Spectrum the symptoms on the middle leaves consisted more of a tip and edge scorch from which irregular dying back developed rapidly. Flowers formed after the onset of the first symptoms were of poor quality. Premature death of the calyx was often seen, the upper margins being first affected. Holley (1956) described leaf burn in the lower foliage and necrotic spots in the middle leaves. The upper leaves of plants were often scorched or had dead spots on them. Yield, grade and keeping quality were all reduced. Winsor and Long (1962) noted a varietal difference in response to K deficiency; whereas Improved William Sim showed a general loss of leaf colour, sometimes associated with tip-burn, Saugus White showed widespread tip-burn with little or no chlorosis. Both types of symptoms were later reported at different positions on two Sim varieties (Winsor et al., 1970). Thus the lower leaves developed tip-burn whilst the uppermost leaves showed sunken white spots, particularly when the flower buds were swelling. Both types of symptoms occurred in the middle foliage. In extreme cases of deficiency the flowers were of poor colour and quality. The symptoms of K deficiency, including tip-burn and pale sunken spots in the leaves, were illustrated by Penningsfeld (1970).

Winsor and Long (1962) grew carnations in a soil-based compost at two concentrations of K (41 and 145 mg l^{-1} K) in the irrigation water. Flower production by cv. Saugus White was increased by 6.1% at the higher K level, accompanied by a more healthy condition of the foliage during the second season of growth. Flowering of a second cultivar, Improved William Sim, was not affected by the higher concentration of K, but the weight of plant material at the end of the trial showed positive responses to K by both varieties (19% and 8% respectively for the Saugus and Sim varieties). Splitting of the calyx, averaged for both cultivars, was increased by 20.9% at the higher level of K. Three concentrations 25, 108 and 191 mg l^{-1} K were included in a further factorial study of carnation nutrition in a soil-based compost (Winsor et al., 1970). Flower production per unit area was decreased at the lowest K level by 5.6% (P <0.01) and 20.1% (P <0.001) respectively in two sucessive crops, in which two cultivars (William Sim and Pink Sim) were both grown for 22 months. When plots deficient in N and P were omitted from the 96 combinations of nutritional treatments, low K decreased flower production by 9.2 and 26.1% respectively in the first and second crops. A concentration of 108 mg l^{-1} K in the nutrient solutions was sufficient to avoid symptoms of K deficiency in these trials and little further reponse in terms of flower numbers was found at 191 mg l^{-1} K. No precise optimum for liquid feeding could be deduced from the data, but an intermediate concentration of 166 mg l^{-1} K (200 mg l^{-1} K_2O) was considered safe and suitable. Many American workers have advocated supplying 200

mg l^{-1} K in the irrigation water for commercial carnation production (see, for example, Gianfagna and Nichols, 1960; Caldwell and Kiplinger, 1960, apparently based on work published subsequently by Eck et al. (1962)). High K (191 mg l^{-1} K) has been shown to increase splitting of the calyx slightly but significantly (P <0.05; Winsor et al., 1970), confirming the work of Blanc (1968).

Growth responses. Potassium chloride is a less expensive source of K than potassium nitrate, and many U.S. growers have been reported to use it for carnation production (Bond and Hartley, 1978). Inclusion of chloride at 3 meq l^{-1} (107 mg l^{-1}) in the nutrient solutions applied to soil-grown carnations reduced growth (fresh weight) by 20% (Hartman and Holley, 1968), and the authors advised against the use of potassium chloride by commercial carnation growers. Bond and Hartley (1978) compared the growth and flowering of carnations supplied with K as either nitrate or chloride with plants grown in soil and in gravel. Potassium chloride caused shortening of the internodes and brittleness of the stems, particularly in gravel culture, and many buds turned brown and failed to develop.

Potassium content of the plants. Published data for the K content of carnation leaves show fairly wide variation; the values in Table 22, recommended or found on commercial holdings, range from 2.0 to 6.3% K. Deficiency was indicated below 1.5–2.0% K (Winsor et al., 1970) or below 2% K (Fortney and Wolf, 1981). Part of the variation in levels was associated with the age of the crop, higher values being found in the first than in the second season of growth (Winsor et al., 1970; Kazimirova, 1977). Parker and Holley (1972) showed a rapid decline in K content of the leaves over the period 8–18 weeks from planting. The 'standard tissue concentrations' developed by Nelson and Boodley (1965) indicate preferred values of between 5 and 6% K in the leaves during the first autumn of a June-propagated crop, falling steeply to values of about 3% K thereafter. Criley et al. (1983) reported a decline in leaf K values in field-grown carnations from the first to the third flushes. Dieback of the calyx tips occurred when the calyx contained less than 0.4% K. Hanan (1974) recorded that, out of total of 39 greenhouses studied, K deficiency had been found in no less than 38% of them. Amongst the 173 leaf samples analysed, 20% contained less than 3.0% K whilst 54% exceeded 3.4% K, these being the approximate limits recommended for carnations in Colorado.

Calcium

Symptoms and growth responses. Clapp and Folley (1941) grew carnations in sand culture with and without Ca. The deficient plants turned pale green and died within 9 weeks of omission of this nutrient, but no clearly defined symptoms of deficiency were found. The first indication of deficiency was a very characteristic form of tip-burn of the leaves (Messing and Owen, 1952; Messing, 1953, 1955, 1958). A constriction developed some distance from the tip and the affected tip bent upwards at an angle to the rest of the leaf; see illustration by Messing (1955). The youngest leaves were the most affected, and death of the growing points followed. Numerous tall, thin, sideshoots developed subsequently and in turn became affected by tip-burn. Under the extreme conditions of sand culture, necrotic spots developed on the stem above an upper pair of leaves, leading to collapse of the stem at that point. This condition subsequently developed above successive nodes down the stem.

Holley and Baker (1963) noted that, whilst symptoms of deficiency were extremely rare on soil-grown plants, the deficiency was not at all uncommon where plants were grown in soilless substrates. A tip-burn similar to that found by Messing (1958) was described and illustrated by Peterson (1960a). In addition, the flowers tended to develop enlarged and protruding pistils, this being considered a sign of premature ageing, and some flowers failed to open. The base of the stems became enlarged, producing a '. . . gall-like overgrowth similar to crown gall or clubroot of cabbage' (Holley and Baker, 1963). Hanan (1975) referred to Ca deficiency in terms of ' . . . typical necrosis of the leaf tips during high light periods'. Tip-burn of carnation leaves was also associated with deficiency by Puustjärvi (1976); spraying with calcium nitrate decreased the incidence of tip-burn, but did not eliminate it entirely. Berghoef and Elzinga (1982) reduced the incidence of leaf scorch by spraying with 1% calcium chloride or 1.6% calcium nitrate.

The vase-life of cut flowers of cv. Red Sim was reduced by low Ca without affecting yield or growth (Holley, 1956). Chan et al. (1958) grew carnations (cv. William Sim) in an inert medium (Haydite) at concentrations of 50, 150 and 300 mg l^{-1} Ca in the nutrient solution; no significant response to Ca was found. Hartman and Holley (1968) found that 2.5 meq l^{-1} Ca (50 mg l^{-1}) was insufficient to maintain the required level in the tissue (1% Ca or above); a concentration of 6 meq l^{-1} (120 mg l^{-1}) was recommended for plants grown in soil or perlite. Whilst the responses of carnations to calcium carbonate are doubtless due as much or more to effects of pH as of Ca supply, it is convenient to include references to liming under this heading. Peterson (1960a) grew White Sim carnation in volcanic scoria with three rates of liming (0, 5 and 10 lb calcium carbonate per 100 sq ft, equivalent to 0, 0.245 and 0.491 kg m^{-2}). Flower production was decreased by 13% where calcium carbonate was omitted. Increased incidence of calyx splitting in limed soil in the absence of added B was also reported by Adams et al., 1979; the interaction between liming and B supply was significant at P <0.001.

Liming the soil increased splitting of the calyx from 11.8% to 16.3% (P < 0.001; Winsor et al., 1970), this effect being attributed to decreased availability of B in soils at high pH. Leaves from the limed plots contained only 15 μg g^{-1} B or less as compared with 20–40 μg g^{-1} where lime was omitted. In a second experiment in the same plots with B included in the irrigation water, liming no longer affected the incidence of calyx splitting significantly.

Calcium content of the leaves. Hanan (1975) planted successive test crops of carnations (cv. White 'Pikes Peak') at two month intervals, and harvested them when the first flowers became visible. The Ca and Mg contents of the leaves increased during periods of low light. At high radiation levels, Ca concentrations below 1% in the leaves resulted in tip-burn. During the winter, however, no deficiency symptoms developed even at Ca levels down to 0.65% or below. The Ca content of the leaves of carnations grown in gravel decreased progressively with increasing concentrations of total salts in the nutrient solutions (r = −0.86; 8 df). The levels recorded in Table 22 are generally within the range 1–2% Ca in the leaf dry matter. Fortney and Wolf (1981) regarded values below 0.60% as deficient. The value of 0.43% Ca reported by Puustjärvi (1972) from experiments in Finland is considerably below the general range of findings; it is possible that such values are associated with low radiation levels and restricted transpiration during the winter months at northerly latitudes. Far higher values for Ca (1.4-1.9%) were subsequently recommended by Puustjärvi (1981b).

Magnesium

Symptoms and growth responses. The symptoms of Mg deficiency were described by Messing and Owen (1952), Messing (1953, 1955, 1958) for three cultivars (White Maytime, Spectrum and William Sim) grown in sand culture. Chlorosis appeared suddenly in the middle leaves and spread rapidly up and down the plant. Much of the foliage became virtually yellow, with only the veins remaining light green. The stems were weak, growth was checked and the old leaves died rapidly. Messing and Owen (1952) also described a very characteristic symptom which occurred at a later stage, consisting of a narrow, necrotic band across the leaf blade, close to its base. The affected leaves collapsed, with up to six consecutive pairs of leaves hanging down; for illustration, see Messing (1955). A close similarity was noted between this basal banding pattern and the effects of Mg deficiency in Easter Lilies (Seeley, 1950).

Reports of the responses of carnations to Mg other than in sand culture are rare. Holley and Baker (1963) included recommendations for supplying magnesium either in solution at 12.5 mg l^{-1} or, where necessary, to the soil at 1 lb magnesium sulphate per 100 square feet ≡ 49 g m^{-2}.

They noted, however, that this element was usually present in the water supply and in limestone incorporated in the substrate. An unpublished study by R. C. Peterson was also referred to by Holley and Baker (1963: p. 35) in which carnations (cv. Pink Sim) were grown in scoria for two years. No visual symptoms of deficiency were observed on plants not receiving Mg, nor were yield or growth responses obtained. Magnesium sulphate $(MgSO_47H_20)$ was included at two levels (0 and 1.78 kg m^{-3}) in a factorial nutritional study in a soil-based substrate (Winsor and Long, 1962). Responses in this trial were minimal, including a slight increase in the proportion of small, 'second quality', blooms (P < 0.01) where Mg had been added. Parker and Holley (1972) found no effect on yield at 1 meq l^{-1} Mg compared with 3.5 meq l^{-1} Mg in the 'standard' treatment, but reduced concentrations were evident in the leaves as early as eight weeks after planting (0.16% Mg compared with 0.31% in the 'standard' plants). Four different ratios of Ca to Mg in the irrigation water were compared by Hartman and Holley (1968), the combined concentration of these two nutrients being maintained at 7 meq l^{-1}. A concentration of 3.5 meq l^{-1} Mg (43 mg l^{-1} Mg) sufficed for maximum growth (fresh weight) of carnations in both soil and perlite, the corresponding Mg contents of the leaves being 0.54% and 0.40%.

Magnesium content of the leaves. Published analyses of carnation leaves (Table 22) show a fairly wide range from 0.2% to 0.6% Mg, probably reflecting the rather limited responses observed to this nutrient. Most of the analyses of carnation leaves reported for experimental crops by Hartman and Holley (1968) were in the range 0.20–0.31% Mg. Values down to 0.11% Mg were reported in plants grown at a low concentration (4.9 mg l^{-1} Mg) in an inert substrate (perlite). A higher concentration (61 mg l^{-1} Mg) gave a leaf content of 0.74% Mg. Their recommended fertiliser treatment was intended to give a content of not less than 0.35% Mg. In one experiment, maximum growth in soil and in perlite corresponded to 0.40% Mg and 0.54% Mg respectively in the leaves. Parker and Holley (1972) reported leaf values increasing slightly with age from 0.31% after treatment at 85 mg l^{-1} Mg for 8 weeks to 0.39% Mg after 18 weeks. The corresponding values in the 'low Mg' treatment (24 mg l^{-1} Mg) were 0.16% and 0.21% Mg, without significant reduction in yield (total plant weight). Bond and Hartley (1978), comparing K sources for carnations, recorded 0.27–0.36% Mg in the leaves of plants grown in gravel culture compared with 0.34–0.36% Mg for plants grown in the soil. No deficiency symptoms were found in leaves containing 0.15% Mg (Adams and Winsor, unpublished data), and deficiency was suggested by Fortney and Wolf (1981) at leaf concentrations below this level. Peterson (1982) classed 0.16–0.24% Mg as low, 0.25–0.50% as 'sufficient' and 0.51–0.55% as high; concentrations above 0.56% Mg were regarded as excessive.

Iron

Messing (1953), who recorded the deficiency symptoms associated with most of the essential elements, was unable to induce visual symptoms of Fe deficiency in carnations grown in sand culture. His data, however, show reductions of 22% and 17% in the numbers of blooms per plant from cvs. White Maytime and Spectrum respectively when deprived of Fe. His early results also suggested that this deficiency had a detrimental effect on splitting of the calyx. Closer study of the data, however, revealed marked differences between individual plants (Messing, 1958). By propagating from these plants and comparing the effects of Fe deficient and complete nutrient solutions in the new generation it was established that the tendency to calyx splitting was inherited and was independent of Fe nutrition. Reduced flower production by plants deprived of Fe was, however, confirmed for both 'splitting' and 'non-splitting' plants.

Peterson (1960b) grew plants of cv. Pink Sim in an inert substrate (volcanic scoria) irrigated with nutrient solutions containing 0, 3 and 6 mg l^{-1} Fe. Yield, stem length and weight per bloom were all unaffected by the treatments, and it was concluded that carnations were not sensitive to Fe concentration in the nutrient solutions. As in Messing's early studies (1953), the highest incidence of splitting occurred in the deficient plants, but this was attributed by Peterson to differences in position on the glasshouse bench. Holley and Baker (1963), noting the work of both Messing (1955) and Peterson (1960b), concluded that Fe should rarely become a limiting factor in the culture of carnations. The orange and yellow-flowered varieties were noted as particularly prone to show Fe chlorosis, however, and this was especially a problem in the highly calcareous soils along the Italian Riviera. This condition could be corrected by inclusion of chelated Fe in the irrigation water at the rate of 1 ounce per 1000 USA gallons (7.5 mg l^{-1}).

Carnations grown in gravel culture (Parker, 1971a) showed the greatest production of plant material (fresh weight) when chelated Fe (Fe–DTPA) was supplied in solution at 3 mg l^{-1} Fe. Omission decreased growth slightly (5.3%) but not significantly, whereas high levels (5, 15 and 50 mg l^{-1} Fe) depressed growth significantly (P < 0.05). The Fe content of the leaves (49–61 μg g^{-1} Fe) did not increase appreciably over the range 0–15 mg l^{-1} Fe, despite the adverse effects on growth at 5 and 15 mg l^{-1}, but rose to 170 μg g^{-1} at the highest concentration tested (50 mg l^{-1} Fe in the solution). Parker (1971a) recommended levels of 50–100 μg g^{-1} Fe in the leaves as adequate, with deficiency in the range 30–50 μg g^{-1} Fe; see Table 23. Other leaf analyses shown in Table 23 include mean values of 73 μg g^{-1} (White, 1966) and 80 μg g^{-1} (Puustjärvi, 1972). Boodley, quoted by Holley (1968), regarded 50–150 μg g^{-1} Fe in the leaves as desirable, whilst Hanan (1975) recommended 50–100 μg g^{-1} Fe and Criley and Carlson

(1970) quoted an optimal value of 100 $\mu g\ g^{-1}$ Fe. Fortney and Wolf (1981) indicated deficiency below 30 $\mu g\ g^{-1}$ Fe, with 50–150 $\mu g\ g^{-1}$ as normal. Peterson (1982) regarded 50–150 $\mu g\ g^{-1}$ Fe as 'sufficient', 31–49 $\mu g\ g^{-1}$ as low and values up to 30 $\mu g\ g^{-1}$ as deficient.

Manganese

Messing (1953, 1955, 1958) observed no visual symptoms of Mn deficiency when this nutrient was omitted in sand culture. Since acute symptoms of this deficiency had earlier been produced in chrysanthemums and tomatoes, it was concluded that the requirement by carnations was relatively low. Attempts to show Mn deficiency in carnations of Colorado State University also proved unsuccessful (Holley and Baker, 1963).

Manganese contents found or recommended in the leaves of normal carnation plants (see Table 23) include 30–250 $\mu g\ g^{-1}$ (Puustjärvi, 1981a), 50–150 $\mu g\ g^{-1}$ (Parker, 1971a; Hanan, 1975) and 100–300 $\mu g\ g^{-1}$ (Boodley, quoted by Holley, 1968; Fortney and Wolf, 1981; Peterson, 1982). Values of 20–50 $\mu g\ g^{-1}$ Mn (Parker, 1971a) and 31–99 $\mu g\ g^{-1}$ (Peterson) were regarded as low. Deficiency was indicated below 30 $\mu g\ g^{-1}$ Mn (Fortney and Wolf, 1981; Peterson, 1982).

Penningsfeld (1970) reported 49–55 $\mu g\ g^{-1}$ Mn in the flower stalks (pedicels) of healthy carnations, with deficiency at 16 $\mu g\ g^{-1}$ Mn.

Manganese toxicity. At the other end of the concentration range, Mn toxicity in carnations has attracted much attention, the crop having proved somewhat susceptible to the high levels frequently encountered in steamed soils. Coorts (1958) grew carnations in nutrient solutions containing up to 200 mg l^{-1} Mn; the visual symptoms included tip-burn, necrotic spotting and distortion of the leaves. The same general symptoms were also produced in soil-grown plants, except that tip-burn of the leaves was more severe. It was reported by White (1967, 1971) that carnation growers in Pennsylvania, USA, had found appreciable numbers of stunted plants in their production benches. This stunting became noticeable some 6–10 weeks after planting into steam sterilised soil. At later stages of growth a tip-burn developed on the leaves, progressing down the mid-vein on the lower foliage. The affected leaves contained 1300–5500 $\mu g\ g^{-1}$ Mn as compared with <500 $\mu g\ g^{-1}$ in plants not showing toxicity symptoms. A factorial experiment was made in which 1 part of either vermiculite or peat moss was added to a 2:1 mixture of soil and perlite. Carnations were grown in these mixtures, both unsteamed and after steaming for 1.5 or 6 hours, with an without addition of manganese sulphate ($MnSO_4.H_2O$ at 165 kg m^{-3}). The Mn content of the leaves increased markedly both with the duration of steaming and with the incorporation of peat moss (pH 5.5). Whilst the incorporation of manganese sulphate did not raise the Mn content of plants grown in unsteamed soil to toxic levels, it greatly increased levels in the leaves of plants in steamed soil. Symptoms of toxicity occurred at concentrations down to 700 $\mu g\ g^{-1}$ Mn, and both yield and quality declined at concentrations of 2250 $\mu g\ g^{-1}$ Mn and above.

According to White (1968), Mn toxicity first appeared as a purplish colour on the tip of the lower leaves, progressing towards the base; the tissue originally affected changed to a straw colour as the injury developed. The plants were described as hard, stiff and stunted. Toxicity occurred above 700 $\mu g\ g^{-1}$ Mn, and levels of 5000 $\mu g\ g^{-1}$ Mn were occasionally reached. These symptoms, noted by White (1968) as '. . . different from those previously described', were to be distinguished from those of B deficiency, where purpling of the leaf tips occurred on the young leaves. White (1971) subsequently recorded that growth of carnations was depressed by high Mn levels.

Parker (1971a) studied the responses of carnations at concentrations of 0, 0.5, 10, 25 and 50 mg l^{-1} Mn in gravel culture. Over a period of six months, plant growth (fresh weight) was not depressed where Mn was omitted from the solutions; the leaves still contained 63 $\mu g\ g^{-1}$ Mn. Manganese at 10 mg l^{-1} did not decrease growth in this relatively short-term experiment, and the content of the leaves remained acceptable (342 $\mu g\ g^{-1}$). Concentrations of 25 and 50 mg l^{-1} Mn induced high levels of tissue Mn (1190 and 1745 $\mu g\ g^{-1}$), however, accompanied by reductions in growth.

The growth and flowering of two carnation cultivars in a high Mn soil at four pH levels, with and without soil sterilisation, were studied by Ishida and Masui (1973). Suppression of growth, tip-burn of the leaves and chlorosis of the young leaves and stems were associated with high levels in the plants. Normal and chlorotic plants of cv. Coral contained 1200 and 3525 $\mu g\ g^{-1}$ Mn respectively in the leaves and 388 and 1075 $\mu g\ g^{-1}$ Mn in the stems; the content reported for the leaves of 'normal' plants in this study was well above those commonly encountered (cf. Table 23). Growth retardation and tip-burn symptoms were highly correlated with the Mn content of the plants and soils, but not with the content of P, K, Mg and Al in the leaves or with NH_4–N, NO_3–N, Na, Mg, P or Al in the soils (Ishida and Masui, 1976). Symptoms of tip-burn on the upper leaves of cv. Coral, grown in sand culture at concentrations of 60, 90, 150 and 300 mg l^{-1} Mn, appeared after 120, 110, 50 and 40 days respectively. Toxicity symptoms occurred at a concentration of 2600 $\mu g\ g^{-1}$ Mn in the leaves (Ishida et al., 1977). Fortney and Wolf (1981) recorded toxicity at far lower levels, down to 800 $\mu g\ g^{-1}$ Mn; see also Peterson (1982).

Copper

Penningsfeld and Forchthammer (1970) described and illustrated the effects of Cu deficiency on carnations. The stems became so thin and weak that they were unable to support the weight of the flowers. Because of this loss of quality the flowers were unsaleable; see also Penningsfeld (1970).

Parker (1971a) grew carnations at five levels of Cu (0, 0.5, 1, 3 and 7 mg l^{-1}) in gravel culture. Omission did not reduce the yields of plant material (fresh weight) during the six-month period of the trial, and no significant differences were found over the range 0–3 mg l^{-1} Cu; the leaves contained 6–14 µg g^{-1} Cu. Copper at 7 mg l^{-1} depressed the yield by 29%, however, accompanied by a sharp rise in leaf content to 50 µg g^{-1} Cu.

Adams *et al.* (1973) reported that, unlike the chrysanthemum, flowering of which was markedly affected by Cu deficiency, carnations appeared to be relatively tolerant in this respect. The yields and grading of carnations grown with and without added Cu at two pH levels in beds of peat were reported subsequently (Adams *et al.*, 1980). Omission reduced the vigour of the plants at high pH, but no specific symptoms were observed, nor any significant effects on either total flower numbers or on grading. Leaf analyses nevertheless showed very low levels in the 'no-copper' plots, namely 1.0 and 0.9 µg g^{-1} Cu at pH 5.9 and 7.2 respectively compared with 7.2 and 5.4 µg g^{-1} Cu where Cu was supplied.

Published data for typical or satisfactory levels in carnation leaves (Table 23) include 5–10 µg g^{-1} (Hanan, 1975), 7 µg g^{-1} (White, 1966), 5–15 µg g^{-1} (Puustjärvi, 1981a), 10–30 µg g^{-1} (Boodley, quoted by Holley, 1968; Fortney and Wolf, 1981; Peterson, 1982) and 20 µg g^{-1} (quoted by Criley and Carlson, 1970; Puustjärvi, 1972). 'Low' levels of Cu were designated by Parker (1971a) as 4–5 µg g^{-1}, and by Hanan (1975) as 4 µg g^{-1}. Fortney and Wolf (1981) and Peterson (1982) regarded deficiency as likely to occur below 5 µg g^{-1} Cu.

Penningsfeld (1970) reported 4.3–4.8 µg g^{-1} Cu in the flower stalks of normal carnations, falling to 1.3 µg g^{-1} Cu under conditions of deficiency.

Zinc

Messing (1985), who described the symptoms of many nutritional deficiencies of carnations grown in sand culture, was unable to demonstrate Zn deficiency. Holley and Baker (1963) confirmed that this also had been the case in experiments at Colorado State University. They suggested that the use of galvanised iron pipe in the water supply to commercial nurseries usually ensured an adequate level of Zn for irrigated crops. Penningsfeld and Forchthammer (1970) also failed to induce visual symptoms of deficiency in carnations grown in a 1:1 mixture of peat and polystyrene granules (Styromull), but a significant depression in yield was noted where Zn was omitted entirely. Parker (1971a) grew carnations in gravel at five concentrations of Zn (0, 0.2, 10, 25 and 50 mg l^{-1}). Total yield (fresh weight) was reduced significantly both where it was omitted and also when supplied at high concentration (25 and 50 mg l^{-1}). A concentration of 0.2 mg l^{-1} Zn was

sufficient to supply the requirement of the plants, though the yield was not depressed significantly at 10 mg l^{-1} Zn. It was suggested that, where chelated Fe compounds such as Fe-DTPA are used (Sequestrene 330), a small amount of Zn should be included to ensure an adequate supply for the crop. Tissue analyses indicated that 21 µg g^{-1} Zn sufficed for maximum growth and that tissue concentrations above 162 µg g^{-1} were excessive.

Further analytical data are included in Table 23. Zn levels regarded as adequate in the leaves include 15–50 µg g^{-1} (Puustjärvi, 1981a), 24–48 µg g^{-1} (White, 1966), 25–75 µg g^{-1} (Boodley, quoted by Holley, 1968; Fortney and Wolf, 1981; Peterson, 1982), 25–100 µg g^{-1} (Parker, 1971a, Hanan, 1975) and 50 µg g^{-1} (quoted by Criley and Carlson, 1970). Penningsfeld reported 32–38 µg g^{-1} Zn in the flower stalks of healthy carnations.

Levels of 18–25 µg g^{-1} Zn in the leaves were classed as 'low' by Parker (1971a), with deficiency indicated below 15 µg g^{-1} Zn (Criley and Carlson, 1970; Fortney and Wolf, 1981; Peterson, 1982).

Boron

Boron deficiency. The symptoms of B deficiency, reported in detail by Messing and Owen (1952) and by Messing (1953, 1955, 1958), included stiffening of the stems and early death of the young flower buds. The axillary shoots often split the base of the leaves, particularly in the Sim varieties; for illustrations see Messing (1955 and 1958). The leaf symptoms differed somewhat between varieties. The young leaves of cv. White Maytime became stiff, remained short and turned yellow at the tip. After some delay, death followed rather suddenly, being preceded by little or no necrosis and tip-burn. In cv. Spectrum there was little loss of colour but an early tip scorch developed. The leaves died back slowly from the tip, and the dried parts became twisted. The surface became covered with small round 'lumps', and red pigment developed in irregular patches. Acute deficiency resulted in characteristic bushy growth, as illustrated by Messing (1955). When B was withheld from mature plants grown in sand culture the flowers which opened at the onset of deficiency (cv. White Maytime) had distorted and transparent edges on the outermost petals. In red varieties the same parts of the petals were almost blackened and the whole flower had an untidy, ragged appearance. Despite the well-marked responses to B deficiency in these studies, Messing and Owen (1952) concluded that splitting '. . . cannot at this stage be related to any particular (nutritional) treatment'.

Holley (1958) described the development of B deficiency symptoms in the following order :- (a) slightly discolouration of the margin of the older leaves, usually bleached along the edges but developing considerable purplish colour in the later stages, (b) reduced numbers of petals in the flowers, (c) extreme malformation of developing flowers and buds and (d) excessive branching of the flower stems. Holley concluded that B was the trace element most likely to become deficient in Colorado greenhouse soils. The customary concentration of 0.5 mg 'boron' (B_2O_3) per litre in the irrigation water (0.16 mg l^{-1} B) was found to be inadequate for carnations grown during the summer in Colorado (Holley, 1959). Inclusion of 0.5 oz borax per USA 1000 gallons of water, equivalent to 0.43 mg l^{-1} B, was therefore recommended to maintain the status of greenhouse soils.

Oertli (1960) described a pale yellow zone along the mid-rib of B deficient leaves, later becoming purple. Boron deficiency was described by Holley and Baker (1963) as common in carnation-growing areas throughout the world. The deficiency first affected the apical tips and buds, followed by excessive lateral branching high on the stems (illustrated). When the glasshouse atmosphere had been enriched with carbon dioxide to promote growth, Goldsberry (1961) found that 1.5–2 mg l^{-1} B were required to avoid deficiency symptoms.

Boron deficiency occurred chiefly in the spring and was eliminated by applications of borax to the soil at the rate of 2 g m^{-2} annually, watered in (Levonen, 1965). The symptoms included splitting of the leaves by axillary shoots, the latter curving through or sometimes growing through the split (cf. Messing, 1958). Early branching of the flower stalks and distorted, spoon-shaped leaves with serrated tips were also reported. Flowering was delayed, dry flower buds developed, and the uppermost lateral shoots sometimes had a 'witches broom' appearance (cf. Holley and Baker, 1963). Symptoms of B deficiency, including shortening of the internodes and deformation of the flowers, were illustrated by Penningsfeld (1970). Heuterman et al. (1973) re-examined the sequence of development of deficiency symptoms in sand culture. The first indication was a shortening of the internodes below the bud, seen 7 months after planting. The tips of the upper leaves turned purple. The stems were stiff and dis-budding became more difficult. The buds stopped growing and dried up. Such flowers as did develop without added B had prominent styles, and none was of commercial quality. Foliar symptoms, including a pale yellow colour along the mid-vein, later becoming purple and finally necrotic, were apparent 10 months after planting. In severe cases of deficiency a 'witches broom' condition developed at the lower nodes; when this occurred during the summer months the plants frequently died and the root systems were found to be poorly developed. Unlike Messing (1958), Levonen (1965) and others, Heuterman et al. (1973) observed splitting of the bases of leaves at all levels tested (0–2.5 mg l^{-1} B) in sand culture. Adams et al. (1980) recorded the sequence of symptom development as follows:- (1) mild deficiency, mainly purpling of the tips of the upper leaves, four weeks after planting, (2) bud failure after six months and (3) yellowing along the centre of the middle and lower leaves some nine months after planting.

Boron toxicity. The amount of B required to correct a deficiency is relatively small, and horticultural advisers have long been aware of the danger of overdosing with the development of a toxic condition in the crop. Lunt et al. (1956) observed a tip-burn of the lower leaves of two Sim cultivars receiving more than 4 mg l^{-1} B in the nutrient solutions. Uniform scorching and death of the leaf tips due to excess were similarly reported by Holley (1959); other symptoms described included intensification of flower colour and an increase in the percentage of split calyces. Eck and Campbell (1962) grew carnations in soil treated periodically at concentrations of 0, 45, 60 and 90 mg l^{-1} B, with and without added limestone; it should be stressed that these concentration, used at intervals of a month or more, are far higher than would be required to induce toxicity symptoms when included regularly in the irrigation water. Necrosis of the tips of the lower leaves appeared after 2–3 months in soil treated at 60 and 90 mg l^{-1} B at monthly intervals. The symptoms occurred first, and were more severe, where the plants were grown in unlimed soil. It was noted that, whilst the symptoms resembled those described for toxicity by Lunt et al. (1956), the excessive branching and production of 'bull-headed' flowers attributed by Holley (1958) to excess B were not observed. Parker (1971b) observed symptoms of toxicity (leaf scorching) in carnations grown in gravel culture with B at 8 mg l^{-1} and Ca at 70 mg l^{-1}; no toxicity symptoms developed, however, when the Ca level was raised to 160 mg l^{-1}, suggesting an interaction between these two nutrients. Parker advised that visual symptoms should not be used to determine the B status of the crop since significant reductions in yield could occur without evidence of obvious symptoms.

Oertli and Kohl (1961) compared the tolerance of 29 plant species, grown in sand culture at a concentration of 10 mg l^{-1} B. The species varied markedly in the time required for toxicity symptoms to occur, ranging from only four days with asters to 77 days for azaleas. Carnations proved to be the second most tolerant plants in the experiment, requiring 57 days to show symptoms of B toxicity. A relationship was suggested between the pattern of toxicity and leaf venation. Thus necrosis occurs at the leaf tip in parallel-veined leaves and in other lanceolate leaves such as the carnation in which the principal vein ends at the tip. Where veins are of more radial distribution, however, as in geranium leaves, toxicity symptoms appear as a zone all around the margin.

Effects of boron on flowering and quality. Lunt et al. (1956) grew carnations (cvs. Red Sim and White Sim) in sand culture at concentrations of 0, 2.0, 4.5 and 10 mg l^{-1} B. Neither the production nor the quality of the blooms was affected, although some tip burn of the leaves occurred at 4.5 and 10 mg l^{-1} B. It was concluded that carnations were very tolerant of boron concentration. Mastalerz and Campbell (1958) found that splitting of the calyx increased as the B content of the soil decreased. Splitting could also be reduced by preventing the night air temperature from falling below 10°C. In the first year of cropping the proportion of split calyces ranged from 23% in the deficient plots to 5–9% in plots where no symptoms of deficiency occurred (Mastalerz, 1958). Splitting increased to 50% in the deficient plots during the second year. Where B deficiency had been confirmed on commercial nurseries it was recommended that 1 oz borax, dissolved in USA 100 gallons of water or nutrient solution, be applied per 400 square feet of bench area i.e. a solution containing 8.5 mg l^{-1} B applied at the rate of 10.2 l m^{-2}. Alternatively, a fritted trace element mixture could be used. It was stressed that factors in addition to B supply may influence splitting, including variety and rapid changes in temperature; ' . . . any factor which causes the calyx to stop growing, even temporarily, can result in splitting'. Holley (1959) found no significant effect of concentration over the range 0.5–10 mg l^{-1} B_2O_3 (0.16–3.11 mg l^{-1} B) on either yield of flowers or the weight of the flower per unit length of stem, the latter being regarded as a quality characteristic; this experiment was, however, designed to test the effects of B toxicity rather than of deficiency.

Eck and Campbell (1962) found no relationship between treatment (0–90 mg l^{-1} B) and the incidence of splitting, though this is hardly surprising since none of the plants could be regarded as deficient; even those grown without added B contained 100–122 μg g^{-1} B in their leaves. Liming the substrate depressed B accumulation in the leaf, calyx and stem tissue. Winsor et al. (1970) grew two Sim cultivars with 96 combinations of macronutrients and lime in two successive 2-year experiments. In the first experiment, liming the soil increased splitting of the calyx from 11.8 to 16.3% (P < 0.001). This response was attributed to lime-induced B deficiency. Thus leaves from the limed plots contained less than 15 μg g^{-1} B compared with 20–40 μg g^{-1} B where lime was omitted. Boron was included in the irrigation water throughout the second experiment at a concentration of 0.5 mg l^{-1} B, however, and liming then had no significant effect in splitting. The overall proportions of split calyces without and with added B in successive experiments were 14.1% and 9.6% respectively. Splitting was not entirely eliminated by B treatment, however, (cf. Messing and Owen, 1952; Mastalerz, 1958). Parker (1971b) suggested that 0.8 mg l^{-1} B in the irrigation water was adequate for carnations in gravel culture under high light conditions, though no adverse effects were noted at 4 mg l^{-1} B. Adams et al. (1973) studied the responses of carnations to nutritional treatments in beds of peat and peat-sand. Omission of B increased calyx splitting. Some of the flower buds failed to develop normally and flower production was decreased appreciably. Heuterman et al. (1973) found no significant differences (P < 0.05) in the numbers of flowers produced in sand culture over the range 0.1–2.5 mg l^{-1} B, though the lowest values occurred at 1–2.5 mg l^{-1} B. More strikingly, however, no blooms of commercial quality were recorded in the absence of added B.

Inclusion of 0.5 mg l^{-1} B in the irrigation water applied to a soil-grown crop increased the total numbers of blooms produced by 4.0% (P < 0.001) and the number of marketable blooms by 58% (P < 0.001). The response to B increased with soil pH; thus application increased the number of marketable blooms per square metre by 11.4% at pH 7.4 compared with only 2.6% at pH 4.7 (Adams et al., 1979). Boron decreased the proportion of blooms with split calyces by 36% overall (P < 0.001), the percentage response ranging from 47% at low N (100 mg l^{-1} N) to 25% at high N level (260 mg l^{-1} N); the N x B interaction was significant at P < 0.05. The relationship between splitting of the calyx and the B content of the leaves is shown in Figure 23; splitting became a serious problem below about 12.5 μg g^{-1} B. Adams et al. (1980) provided further evidence of the effects of B deficiency on carnation flower production and quality (Table 25). Treatments with and without added boron (3.5 g borax m^{-2}) at two pH levels (5.9 and 7.2) were included in a replicated factorial experiment with two carnation cultivars (32 plants of each per plot), grown in a 3:1 peat/sand substrate for 21 months. Boron deficiency reduced the numbers of blooms produced per unit area, particularly in limed peat. The proportion of large blooms with unsplit calyces was decreased by 74%. The percentages of split and unmarketable blooms were greatly increased by deficiency, and 10% of the buds produced failed to develop normally on B deficient plants.

	Complete nutrients	No boron
Total blooms per m^2	724	675
Total blooms per m^2 at pH 7.2	690	567
% large blooms (>6.5 cm): not split	25.1	6.5
% blooms with split calyces	10.1	27.9
% blooms with split calyces in the last 5 months of cropping	5.1	23.6
% unmarketable blooms	3.2	17.2
% bud failure	0	10.0

Table 25. Some effects of boron deficiency on the flowering and quality of carnations in a peat:sand substrate.

Boron content of the leaves. Analytical data relating to the B status of carnations are shown in Table 23. White (1966) recorded 28–53 μg g^{-1} B in carnation leaves (oven dried), with a mean of 37 μg g^{-1} B (Table 23). Boodley, quoted by Holley (1968), regarded 25–400 μg g^{-1} B, a fairly wide range, as acceptable. Hanan (1975) recommended 25–100 μg g^{-1} B, with 20 μg g^{-1} indicating a low status. Fortney and Wolf (1981; see also Peterson, 1982) regarded 30–100 μg g^{-1} B as normal, with deficiency below 25 μg g^{-1} (Table 23), whilst Puustjärvi (1981a) recom-

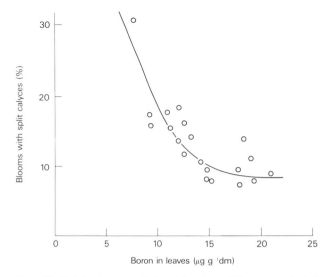

Figure 23 Relation between splitting of the calyx and the boron content of leaves from carnations grown in soil without added boron (after Adams *et al.*, 1979).

mended 30–60 μg g^{-1} B for healthy plants. Further data not included in the Table are also available. Thus Oertli (1960) noted that tissue suffering from deficiency often contained less than 10 μg g^{-1} B, and that where toxicity occurred there was an accumulation of 3000–5000 μg g^{-1} B. Oertli and Kohl (1961) grew a wide range of plants in sand culture at an excessive level of B (10 mg l^{-1} B) and determined the B content of the leaves when toxicity symptoms appeared. Whilst the green parts of the leaves contained 60–680 μg g^{-1} B, the chlorotic and necrotic areas contained 1620–2200 and 1510–5770 μg g^{-1} B respectively. The possibility that the necrotic areas act as sinks and thus reduce the spread of the damage was considered. The various species tested differed more in the time taken to show toxicity symptoms than in the concentration at which such symptoms appeared; the rate at which B accumulated was thus a major factor rather than the tolerance of individual species. Nelson and Boodley (1965a) concluded that the 'lower threshold limit' for deficiency symptoms lay between 20 and 25 μg g^{-1} B in the leaves. The upper limit could not be ascertained accurately from their experiment, but toxicity symptoms first appeared when the content of the leaves reached 420 μg g^{-1} B. Parker (1971b), who grew carnations at four levels (0, 0.8, 4 and 8 mg l^{-1} B) in factorial combination with two levels of Ca (70 and 160 mg l^{-1}), obtained some evidence of a Ca x B interaction at deficiency levels of the micronutrient. Thus plants grown without B at 160 mg l^{-1} Ca contained only 14 μg g^{-1} B in their leaves, a level which would normally be regarded as deficient. At the lower concentration of Ca (70 mg l^{-1}), however, the corresponding B content was 36 μg g^{-1} B, a level which would normally be regarded as fully adequate. No visual symptoms of B deficiency other then stunting were observed at either Ca level.

Heuterman *et al.* (1973) determined the B content of carnation stems, leaves and various parts of the flowers when grown at a range of concentrations (0–2.5 mg l^{-1} B) in sand culture. The leaves showed a progressive increase from 9 μg g^{-1} B in the absence of added B to 347 μg g^{-1} at 2.5 mg l^{-1} B. The calyx and receptacle also showed significant though far smaller increases of B, whilst the stems failed to show significant differences. The remaining plant parts analysed, namely the stigma and style, the ovary wall and the ovary contents, showed no effect of treatment. Since B accumulated to a greater extent in the leaves than in the other plant tissues, especially when the supply was not limiting, leaf analysis was seen as a sensitive indicator of the status of the crop. Adams *et al.* (1979) found no symptoms of deficiency above 21 μg g^{-1} B in the leaves, whereas symptoms were readily apparent below about 15 μg g^{-1} B; the relationship is illustrated in Figure 24. Adams *et al.* (1980) later reported from another experiment that, although symptoms of deficiency were not found above 30 μg g^{-1} B in the leaves, the proportion of flowers with split calyces declined with increasing B status up to 50–60 μg g^{-1} B in the leaves. The percentage of split calyces (y) was related to the B content of the leaves (x) by a quadratic regression of the form:

$$y = 21.8 - 0.46x + 0.0037x^2$$

The regression accounted for 69% of the variance (P < 0.001).

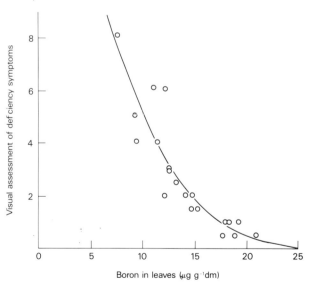

Figure 24 Relation between the incidence of boron-deficiency symptoms on carnations grown in soil without added boron and the boron content of the leaves (after Adams *et al.*, 1979).

Molybdenum

Little information is available concerning the responses of carnations to Mo. Penningsfeld and Forchthammer (1970) grew carnations (cv. Imp. William Sim) at six levels (0,

0.4, 0.8, 1.6, 4.0 and 8.0 mg Mo per litre of peat: polystyrene substrate). Symptoms of Mo deficiency developed 48 weeks after starting the treatments, and a significant depression of growth was recorded where this element was omitted entirely. Penningsfeld (1970) described and illustrated whitish tips on the older leaves of the side shoots, accompanied by very poor quality flowers. Satisfactory growth was found when the flower stalks contained 1.2–1.3 $\mu g\ g^{-1}$ Mo in the dry matter with deficiency at 0.08 $\mu g\ g^{-1}$ Mo. Puustjärvi (1981a) regarded 0.5 $\mu g\ g^{-1}$ Mo in the leaves as normal for healthy carnation plants.

References

Adams P., Graves, C. J. and Winsor, G. W. (1973). Nutrition of carnations *Dianthus caryophyllus* L. in peat and peat-sand. *Rep Glasshouse Crops Res Inst 1972*, 60–61.

Adams, P., Graves, C. J. and Winsor, G. W. (1980). Some effects of micronutrients and liming on the production and quality of glasshouse carnations grown in a peat-sand substrate. *J hort Sci* **55**, 89–96.

Adams, P., Hart, B. M. A. and Winsor, G. W. (1979). Some effects of boron, nitrogen and liming on the bloom production and quality of glasshouse carnations. *J hort Sci* **54**, 149–154.

Beach, G. (1952). Effect of ammonium sulphate and potassium chloride on Patrician carnations in soil. *Proc Am Soc hort Sci* **59**, 484–486.

Berghoef, J. and Elzinga, P. (1982). Calcium chloride vermindert bladverbranding bij 'Pirate'. *Vakblad voor de Bloemisterij* **37** (11), 32–33.

Blanc, D. (1968). Influence du potassium sur la qualite de l'oeillet 'Sim'. *Proc 8th Congr. Int Potash Inst Brussels (1966)*, 177–183.

Bond, K. and Hartley, D. E. (1978). Potassium fertilizer sources for carnations. *Colo Flow Gro Assn Bull* **339**, 1–4.

Caldwell, J. L. and Kiplinger, D. C. (1960). Fertiliser injection. *Carnation Craft* No. **52**, 1–3.

Chan. A. P., Heeney, H. B., Maginnes, E. A and Cannon, H. B. (1958). Mineral nutritional studies on carnation (*Dianthus caryophyllus*) I. Effects of N, P, K, Ca and temperature on flower production. *Proc Am Soc hort Sci* **72**, 473–476.

Clapp, R. and Folley, G. E. (1941). Nutritional symptoms in the carnation. *Proc Am Soc hort Sci* **38**, 673–678.

Coorts, G. D. (1958). Excess manganese nutrition of ornamental plants. *Bull Mo agric Expt Stn* **669**, 35 pp.

Criley, R. A. and Carlson, W. H. (1970). Tissue analysis standards for various floricultural crops. *Florists' Rev* **146**, 19–20, 70–73.

Criley, R. A., Parvin, P. E., Hori, T. M. and Leonhardt, K. W. (1983). Carnation calyx tip dieback. *Acta Hort* **141**, 125–131. *Hort Abs* (1984), **54**, No. 5509.

Eck, P. and Campbell, F. J. (1962). Effect of high calcium application on boron tolerance of carnation, *Dianthus caryophyllus*. *J Am Soc hort Sci* **81**, 510–517.

Eck, P., Campbell, F. J. and Spelman, A. F. (1962). Effect of nitrogen and potassium fertiliser treatment on soil concentrations and on production, quality and mineral composition of carnation. *Proc Am Soc hort Sci* **80**, 565–570.

Fortney, W. R. and Wolf, T. K. (1981). Plant Analysis. *Penn Flo Gro Bull* **331**, 1, 5–11.

Gianfagna, A. and Nicholas, L. P. (1960). Commercial carnation production. *Carnation Craft* No **53**, 2–5.

Goldsberry, K. L. (1961). Effects of carbon dioxide on carnation growth. *Colo Flow Gro Assn Bull* **138**, 6 pp.

Green, J. L. and Holley, W. D. (1974). Effect of the NH_4^+ :NO_3^- ratio on net photosynthesis of carnation. *J Am Soc hort Sci* **99**, 420–424.

Hanan, J. J. (1974). Carnation tissue analyses. *Colo Flow Gro Assn Bull* **284**, 4 pp.

Hanan, J. J (1975). Calcium tissue levels in the carnation. *Colo Flow Gro Assn Bull* **305**, 1–4.

Hartman, L. D. and Holley, W. D. (1968). Carnation nutrition. *Colo Flow Gro Assn Bull* **221**, 8 pp.

Heuterman, G. A., Woodbridge, C. G. and Kalin, E. W. (1973). Boron levels in stem, leaves and flower parts of carnation. *HortSci* **8**, 126–127.

Holley, W. D. (1956). Nutritional control for carnations. *Maryld Florist No. 31*, 1–7.

Holley, W. D. (1958). Trace elements nutrition of carnations. *Colo Flow Gro Assn Bull* **105**, 4 pp.

Holley, W. D. (1959). Boron tolerance by carnations and symptoms of excess boron. *Colo Flow Gro Assn Bull* **113**, 4 pp.

Holley, W. D. (1968). Nutrient levels in tissue of greenhouse crops. *Colo Flow Gro Assn Bull* **220**, 4 pp.

Holley, W. D. and Baker, R. (1963). Carnation Production. *Dubuque, Wm. C. Brown Co. Inc.*

Ishida, A. and Masui, M. (1973). Studies of the manganese excess of carnation I. The steam sterilisation and pH levels of soil. *J Jap Soc hort Sci 42*, 40–48.

Ishida, A. and Masui, M (1976). Studies of the manganese excess of carnation II. Manganese excess in relation to certain soils, soil pH and two nitrogen forms. *J Jap Soc hort Sci* **45**, 283–288.

Ishida, A., Masui, M., Ogura, T. and Nukaya, A. (1977). Studies on manganese excess in carnation III. Effects of manganese concentration in nutrient solution on growth and flowering. *J Jap Soc hort Sci* **45**, 383–388.

Kazimirova, R. N. (1977). (Leaf analysis for determining the nutrient requirement of carnations). *Trudy gos nikit bot Sada* **71**, 79–91: *Hort Abstr* (1979), **49**, 378.

Khattab, M., Kaufmann, H-G. and Borner, R. (1977). Ergebnisse methodischer Untersuchungen zur Blattanalyse bei Edelnelke zur Ermittlung der Hauptnährstoffversorgung in den unterschiedlichen Entwicklungsstadien. *Arch Gartenb* **25**, 289–304.

Langhans, R. W. (Ed.: 1961). Carnations : a manual of the culture, insects and diseases and economics of carnations. 107 pp. *Cornell University N Y.*

Levonen, H. J. (1965). Bormangel-symptome bei Edelnelken. *Dtsche Gartnerbörse* **65**, 269: *Hort Abstr* 1965, **35**, 873.

Lunt, O. R., Kohl, H. C. and Kofranek. A. M. (1956). Tolerance of carnations to saline conditions and boron. *Carnation Craft* no. **36**, 5–6.

Mastalerz, J. W. (1958). Carnation splitting and boron deficiency. *Penn Flow Gro Bull* **85**, 1–2.

Mastalerz, E. and Campbell, F. (1958). Carnation splitting corrected by boron. *Florogram*, reprinted in *Flor Exch* **130**, 50: *Hort Abstr* 1958, **28**, 453.

Messing, J. H. L. (1953). The effects of some acute mineral deficiencies on perpetual flowering carnations. *Rep Exp Res Stn Cheshunt 1952*, 71–74.

Messing, J. H. L. (1955). The visual symptoms of some mineral deficiencies on perpetual flowering carnations. *Carnation Craft* No **32**, 1–7.

Messing, J. H. L. (1958). Mineral nutrition of carnations. *J Sci Fd Agric* **9**, 228–234.

Messing, J. H. L. and Owen, O. (1952). The effects of some acute mineral deficiencies on perpetual flowering carnations. *Rep Exp Res Stn Cheshunt 1951*, 78–81.

Ministry of Agriculture, Fisheries and Food (1967). A manual of carnation production. *Minist Agric Fish Fd Bull* **151** (3rd ed). 158 pp. HMSO Lond.

Ministry of Agriculture, Fisheries and Food (1982). Lime and fertiliser recommendations. No. 4 Glasshouse crops and nursery stock 1983/84. *Booklet* **2194**, 48 pp.

Nelson, P. V. and Boodley, J. W. (1963). Selection of a sampling area for tissue analysis of carnation. *Proc Am Soc hort Sci* **83**, 745–752.

Nelson, P. V. and Boodley, J. W. (1965a). Foliar analysis of carnation III: Development of standard tissue concentrations. *N Y State Flow Gro Bull* **230**, 1–4.

Nelson, P. V. and Boodley, J. W. (1965b). Foliar analysis of carnation IV: Classification of carnation varieties according to foliar nutrient content. *N Y State Flow Gro Bull* **234**, 1–6.

Oertli, J. J. (1960). Die Verteilung des Bors in Nelken. *Gartenbauwissenschaft* **25**, 287–292: *Hort Abstr* (1961), **31**, 852.

Oertli, J. J. and Kohl, H. C. (1961). Some consideration about the tolerance of various plant species to excessive supplies of boron. *Soil Sci* **92**, 243–247.

Parker, J. (1971a). Carnation micronutrition. *Colo Flow Gro Assn Bull* **258**, 4 pp.

Parker, J. (1971b). Calcium-boron nutrition of carnations. *Colo Flow Gro Assn Bull* **259**, 4 pp.

Parker, J. B. and Holley, W. D. (1972). The effect of stage plant development on nutrient content of the leaves. *Colo Flow Gro Assn Bull* **261**, 4 pp.

Penningsfeld, F. (1970). 10 Jahre Hydrokulturversuche mit nelken. *Zierpflanzenbau* **10**, 138–142: *Hort Abstr* (1970), **40**, 1054.

Penningsfeld, F. and Forchthammer, L. (1970). Essais pour la détermination de l'approvisionnement optimal en oligoéléments des fleurs a couper. *Proc 6th Colloq Int Pot Inst Florence Italy 1968*, 364–378.

Peterson, R. (1960a). Calcium hunger in carnation. *Colo Flow Gro Assn Bull* **119**, 6 pp.

Peterson, R. (1960b). Iron hunger in carnations should be rare. *Colo Flow Gro Assn Bull* **125**, 4 pp.

Peterson, J. C. (1982). Monitoring and managing nutrition, Part IV: foliar anaylysis. *Ohio Florists' Assn Bull* **632**, 14–16.

Puustjärvi, V. (1972). Nutrient requirements of carnations. *Peat and Pl Yearbook 1971*, 12–13.

Puustjärvi, V. (1976). Potassium/calcium ratio as a regulator of the water consumption of plants. *Peat & Pl Yearbook 1973–1975*, 30–33.

Puustjärvi, V. (1981a). Rationalized micronutrient fertilization. *Peat & Pl Yearbook 1980*, 3–12.

Puustjärvi, V. (1981b). Nutrient level of Finnish peat cultures. *Peat & Pl Yearbook 1980*, 13–17.

Schekel, K. A. (1971). The influence of increased ionic concentrations on carnation growth. *Proc Am Soc hort Sci* **96**, 649–652.

Seeley, J. G. (1950) Mineral nutrient deficiencies and leaf burn of Croft Easter Lilies. *J Am Soc hort Sci* **56**, 439–445.

Szendel, A. J. (1940). Symptoms of potassium deficiency in carnation plants. *Proc Am Soc hort Sci 1939* **37**, 1012.

White, J. W. (1966). Plant analysis for flower crops. *Penn Flow Gro Bull* **187**, 1–4.

White, J. W. (1967). Manganese toxicity. Where do we go from here? *Penn Flow Gro Bull* **191**, 4–5.

White, J. W. (1968). Trace element toxicities for carnations. *Penn Flow Gro Bull* **207**, 1–2.

White, J. W. (1971). Interaction of nitrogenous fertilizers and steam on soil chemicals and carnation growth. *J Am Soc hort Sci* **96**, 134–137.

Winsor, G. W. and Long, M. I. E. (1962). A factorial nutritional study of the perpetual-flowering carnation grown under glass. *J hort Sci* **37**, 299–312.

Winsor, G. W., Long, M. I. E. and Hart, B. M. A. (1970). The nutrition of glasshouse carnations. *J hort Sci* **45**, 401–413.

CHAPTER 7

Chrysanthemum

After the cutting is planted, growth of the chrysanthemum falls into two phases, namely a period of rapid vegetative growth followed by flowering. As the plant changes from the vegetative to the flowering phase, the foliage darkens and the rate of stem elongation declines. Considerable re-distribution of nutrients occurs within the plant at the onset of flowering, with appreciable amounts of nutrients moving into the developing flower buds and upper leaves (Lunt *et al.*, 1964). Bunt (1971) showed that, whilst the total uptakes of N, P and K were four times higher in summer than in winter, the actual concentrations of these nutrients in the dried leaves were considerably lower in summer. Leaf analyses should, therefore, be interpreted in relation to position on the plant, stage of growth and time of year.

Nitrogen

Deficiency symptoms. The first sign of N deficiency is a loss of colour in the older leaves and a reduction in the size of the younger ones. Later, the young leaves become pale green and assume an erect position whilst the older leaves become yellow, then brown, and die prematurely (Messing and Owen, 1954). Purple or red colouration develops on the older leaves of some cultivars. The colour of the bloom may not be affected, but red and bronze colours usually become more intense. Lunt *et al.* (1964) noted that the petioles of N deficient chrysanthemums were stiff, making the leaves stand out at right-angles to the stems. Numerous reddish necrotic spots developed near the margins of the older leaves on severely deficient plants, and also brownish-red lesions on the stems near the leaf axils. The roots of the deficient plants were longer and less branched than those of plants adequately supplied with N. Extensive development of the roots of plants grown at low N levels was also noticed by Massey and Winsor (1974). The deficiency symptoms described by Roorda van Eysinga and Smilde (1980) included restricted growth, with small pale leaves. The older leaves sometimes developed pink or reddish-brown spots and subsequently died. With severe deficiency, flowering was delayed and the size of the flowers reduced.

Growth responses. Lunt and Kofranek (1958) found that plants grown where all the N was applied before planting had higher fresh weights than those receiving part of it in top dressings. They concluded that a slight deficiency

during the first few weeks of growth retarded growth and adversely affected bloom quality; subsequent applications of N did not compensate for the retardation. The importance of a relatively high level of N during the early period of growth was supported by the results of Winsor and Hart (1965) and Butters and Wadsworth (1974).

Growth was depressed by deficiency (Woltz, 1956; Penningsfeld, 1972), increased up to intermediate levels of N, but was depressed by heavy applications (Woltz, 1956; Joiner and Smith, 1962; Massey and Winsor, 1974; Butters and Wadsworth, 1974). Stem length increased progressively with the concentration of N applied in the liquid feed (Winsor and Hart, 1966). In unsterilised peat, N deficiency reduced plant height by 67% but in the following crop, after steaming, the corresponding depression was only 9% (Adams *et al.*, 1976). Ammonium N generally increased the height and dry weight of the plants compared with those receiving only NO_3-N (Tsujita *et al.*, 1975). With pot chrysanthemums, the dry weight was depressed by N deficiency (Penningsfeld, 1972) and generally increased with the quantity applied during spring and early summer (Sharma and Patel, 1978).

Responses to N were much greater in pots, where the rooting volume is limited, than in borders of soil. For example, omission of N from the base fertiliser dressings and the liquid feed reduced plant height by 65% in pots compared with 14% in the border (Massey and Winsor, 1974).

Flowering was generally delayed by low levels of N (Messing and Owen, 1954; Lunt *et al.*, 1964; Winsor and Hart, 1965, 1966; Massey and Winsor, 1974; Machin, 1982) and by excessive levels (Joiner and Smith, 1961, 1962). In winter, high levels delayed the flowering of crops grown in borders (Butters and Wadsworth, 1974) and in pots (Bunt, 1973). The number of buds was depressed by N deficiency (Woltz, 1956; Penningsfeld, 1972) and generally increased with the quantity of N applied (Winsor and Hart, 1966; Butters and Wadsworth, 1974; see also Fig. 25). Butters and Wadsworth (1974) also reported that high N levels caused additional axillary buds to develop on the primary lateral shoots of spray chrysanthemums – a condition known as 'compounding'. A similar response to high N was also observed in trials at the Glasshouse Crops Research Institute (Winsor and Hart, unpublished). These extra buds were late in developing, and the percentage of the total buds which were open at the time of harvesting declined from 90% to 57% with increasing N (Butters and

Wadsworth, 1974). Massey and Winsor (1974) neverthe-less recorded that the number of open blooms per spray increased with the level of N.

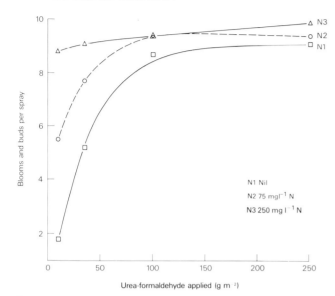

Figure 25 The number of blooms and buds formed on spray chrysanthemums (cv. Hurricane) in relation to the amounts of urea-formaldehyde (38% N) applied to the peat beds before planting and to the concentration of nitrogen in the liquid feed (Adams, Graves and Winsor, unpublished data).

Bloom diameter increased with the applied N level (Winsor and Hart, 1966; Adams *et al.*, 1970; Butters and Wadsworth, 1974; Machin, 1982), though the opposite response was recorded in one crop (Joiner and Smith, 1962). The blooms of plants grown with ammonium sulphate or ammonium nitrate were generally slightly larger than those on plants grown with calcium nitrate only (Tsujita *et al.*, 1974, 1975). In unheated glasshouses, increasing levels of N over the range 0–897 kg ha^{-1} N progressively delayed flowering, reduced bloom diameter and depressed the number of marketable blooms (Carter and Dermott, 1966).

The proportion of large sprays increased with the N level in both the base fertiliser and liquid feed (Adams *et al.*, 1970, 1971, 1975; Massey and Winsor, 1974). After cutting, however, the keeping quality (vase life) of the flowers declined as the N level at which they were grown increased (Joiner and Smith, 1960; Waters, 1965). Nichols (1970) found no effect of N level on the keeping quality of flowers grown in beds of peat, and the results obtained by Butters and Wadsworth (1974) varied from crop to crop. Flowers from plants grown with calcium nitrate lasted longer than those grown with ammonium nitrate (Tsujito *et al.*, 1975).

Nitrogen content of the leaves. The N content of the leaves increased with the N level applied (Woltz, 1956;

Lunt *et al.*, 1964; Waters, 1965; Massey and Winsor, 1974), but this response was relatively small for crops grown in winter (Massey and Winsor, 1974). The N content increased during vegetative growth, but declined after the start of short days (Joiner and Poole, 1967). In sand culture, however, the N status of the foliage generally increased until flower colour developed, the largest increases occurring in the first four weeks after planting (Boodley and Meyer, 1965). The content of the leaves (6.0% N) was unaffected by root temperatures in the range 4.5–16.0°C (Rosenthal *et al.*, 1973), but increased to 6.95% N (P >0.01) at the highest temperature (18–23°C fluctuating). The leaves of plants receiving ammonium sulphate had a higher N content (2.8% N) than those grown at the same level of N with calcium nitrate (2.3%; Tsujito *et al.*, 1974). The total uptake was considerably higher in summer than in winter (Bunt, 1971; Butters and Wadsworth, 1974).

Spray chrysanthemums were reduced in size when the leaves contained only 2.5% N (Butters and Wadsworth, 1974) and 3.1–3.4% N (Massey and Winsor, 1974). The leaves of deficient plants contained 2.25–2.75% N, less than 2% corresponding to severe deficiency (Lunt and Kofranek, 1958), 1.5–3.0% N (Lunt *et al.*, 1964), 1.8% N (Adams *et al.*, 1970), less than 1% (Penningsfeld, 1972) and less than 1% N or less than 0.1% NO$_3$-N (Roorda van Eysinga and Smilde, 1980). For comparison, reported values for healthy plants include 4.0–4.5% N (Lunt *et al.*, 1964), up to 5.6% N (Massey and Winsor, 1974) and 2.5–5.0% N (Roorda van Eysinga and Smilde, 1980).

The following ranges of values have been associated with good growth and flower quality:- 4.5–5.0% N (Woltz, 1956), 4.0–4.5% N (Lunt and Kofranek, 1958), 4.0–5.0% N (Joiner and Smith, 1962), 4.5–6.0% N in the upper leaves (Lunt *et al.*, 1964), 3.5–4.5% N (Waters, 1965), 4.0–5.0% N (Boodley and Meyer, 1965), 5.5% N (quoted by Criley and Carlson, 1970), 4.2–5.9% N (Bunt, 1971), 3.5–5.5% N (Wadsworth and Butters, 1972) and 4.6–5.4% N (Adams *et al.*, 1976). Such values will, of course, vary somewhat with the time of year, since crops grown in good light generally have lower N contents than those grown in winter.

Lunt and Kofranek (1958) reported that leaves grown at high levels of N become very thick and brittle as the plants approached maturity. Affected leaves contained about 5% N. Harbaugh *et al.* (1983) reported increased susceptibility to damage by leafminer (*Liriomyza trifolii*) as the N content of the leaves increased from 2.2% to 4.0 N. Maximum yield of marketable stems was estimated to occur at 3.6% N.

Ammonium toxicity. Ammonia injury occurred on soil (pH 7.2) that had received urea-formaldehyde (50–100 g m^{-2}) after being steamed repeatedly (Massey and Winsor, 1974). The upper and middle leaves of affected plants were pale yellow-green, though the lower leaves remained dark green. The plants tended to wilt in bright sunshine due to the severe root damage and were often made more conspicuous by dark green, stunted plants surrounding them. As growth proceeded, the yellow-green plants usually regained a normal colour, though they were somewhat shorter than healthy plants. Severely affected plants remained permanently stunted.

Symptoms of NH_4-N toxicity, caused by applying high levels of NH_3-N to the soil, have been described by Nelson and Hsieh (1971). After a reduction in the growth rate of the plants, interveinal areas on the middle and lower leaves became yellow. Small dark spots then developed at the base of the laminae of these leaves and spread towards the leaf tip until the whole leaf was affected. The symptoms gradually progressed up the plant and the leaves became abnormally thick and leathery. Root development was greatly reduced, the roots became brown and the plants eventually died. Symptoms were the same in both summer and winter.

These authors found that the symptoms of the toxicity became more severe when the level of NH_4-N was increased or when the level of K applied was decreased. Hence they concluded that the ratio of NH_4-N to K (NH_4-N:K in milliequivalents) in the fresh leaf tissue was a better indicator of NH_4-N assimilation and of the probability of toxicity than the concentration of NH_4-N alone. A critical ratio of 0.025–0.026 was deduced; plants with NH_4-N:K ratios below this value never showed toxicity symptoms whilst those with higher values for the ratio were always damaged.

Phosphorus

Symptoms and growth responses. A deficiency of P reduces the growth rate, and the new leaves formed remain small and erect. The plants become dull green in colour; the oldest leaves become yellow, then brown and wither away. Since these stages succeed one another quite rapidly as they progress up the plant there are only a few yellow leaves on affected plants at any one time. Purplish-red pigments may develop on the leaves of some cultivars (Messing and Owen, 1954). In a similar description, Lunt et al. (1964) noted that the green leaves, which are relatively dark, had a dull greyish hue. The old leaves turned yellow and died. Symptoms of this deficiency, including yellowing and death of the lower leaves, became less severe as the level of N increased (Adams et al., 1970).

The fresh weight of the plants was depressed by P deficiency (Woltz, 1956; Penningsfeld, 1972) and generally increased with the amount applied (Woltz, 1956), but was depressed by heavy applications in one cultivar (Waters, 1964). In most cases the height of the plants increased with the level of P (Winsor and Hart, 1965; Winsor and Hart, 1966; Adams and Winsor, 1967; Joiner, 1967). Massey and Winsor (1969) noted that marked growth responses to P

were not greatly affected by liming, and that responses were smallest in winter.

Phosphorus deficiency delayed flowering (Messing and Owen, 1954; Lunt et al., 1964; Adams, 1975), reduced the number of buds formed (Woltz, 1956; Adams, 1975) and the size of the blooms (Messing and Owen, 1954; Winsor and Hart, 1965; Joiner, 1967; Adams et al., 1970; Bunt, 1971), though P level had little effect on the number and size of the blooms of one cultivar ('Elegance') which flowered in winter (Winsor and Hart, 1966). The proportion of large sprays increased with the rate of P applied (Adams et al., 1971). The colour of the blooms was usually slightly paler on deficient plants (Messing and Owen, 1954).

Phosphorus content of the leaves. The P content of the leaves increased with the level of P applied (Woltz, 1956; Lunt et al., 1964; Waters, 1964; Joiner, 1967) and was higher in the upper than in the lower leaves (Lunt et al., 1964). Pot-grown chrysanthemums had a higher content in their leaves during winter (0.47–0.58% P) than in summer (0.28–0.33% P; Bunt, 1971). The percentage increased during early vegetative growth, declined sharply when short-day treatments were initiated and then increased again during flower development (Joiner and Poole, 1967). Potassium deficiency increased the P content of the leaves (Holcomb and White, 1974; see also Joiner and Smith, 1962). Uptake of P was depressed by reducing the root temperature; this response was particularly marked between 16° and 10°C (Rosenthal et al., 1973). The total uptake of P increased with the application rate and was only slightly depressed when the pH of the soil was increased by liming from 6.1 to 7.5 (Adams and Winsor, 1967).

The leaves of deficient plants contained 0.18% P or less, 0.14% P corresponding to severe deficiency (Waters, 1964), 0.10–0.21% P (Lunt et al., 1964), 0.10% P (Butters and Wadsworth, 1974) and 0.19% P (Roorda van Eysinga and Smilde, 1980), compared with more normal values of 0.30–0.40% P (Woltz, 1956), 0.25–0.75% P (Boodley and Meyer, 1965), 0.52–1.64% P (Joiner, 1967), 0.5% P (Butters and Wadsworth, 1974) and 0.31–0.93% P (Roorda van Eysinga and Smilde, 1980). Critical levels of 0.26% and 0.17% P were suggested for the upper and lower leaves respectively (Lunt et al., 1964). Excessive levels (2.3–2.6% P) were found in the leaves of chrysanthemums which had received heavy dressings of magnesium ammonium phosphate (Sharma and Patel, 1978).

Potassium

Symptoms and growth responses. The symptoms of K deficiency include browning of the margins and death of the lower leaves (Woltz, 1956), but vary somewhat with the cultivar and season (Messing and Owen, 1954). In some cases a yellow mottling appears on the middle leaves. The spots enlarge, becoming brown in the middle, and the leaf margins curl downwards and inwards. The unaffected leaves are slightly darker than those on healthy plants (Messing and Owen, 1949). More generally, a yellow margin, which rapidly becomes brown, develops on the middle leaves. The rate of growth becomes slower, the leaves are greatly reduced in size and the internodes are shortened. The oldest leaves die prematurely, but later than the leaves which first showed the symptoms, and some cultivars produce basal cuttings. A poor root system is characteristic (Messing and Owen, 1954). Lunt and Kofranek (1958) noted that a severe reduction in leaf size was the first symptom of the deficiency. As browning and death of the leaves progressed up the plant, there were only a few leaves between those that were green and the completely brown ones. Winsor and Hart (1965) also reported that deficiency caused death of the lower leaves, except when growth was already restricted by P deficiency. Winsor (1968) noted that the blooms of deficient chrysanthemums (two yellow-orange cultivars) grown in the field were paler than those from adequately fertilised plants, and in extreme cases failed to develop normally. Dead brown areas developed irregularly around the margins of the leaves, the uppermost leaves being the most severely affected. Roorda van Eysinga and Smilde (1980) referred to small (1–2 mm) chlorotic spots, later coalescing, which appeared at the margins of the leaves and later spread interveinally. Growth was thin and feeble.

Potassium deficiency reduced plant height slightly (Lunt and Kofranek, 1958; Lunt et al., 1964; Adams et al., 1970; Holcomb and White, 1974; Massey and Winsor, 1974) and resulted in weak and spindly stems (Lunt et al., 1964). High levels of K also depressed plant height (Winsor and Hart, 1965; Butters and Wadsworth, 1974), particularly at low levels of P (Joiner, 1967). Plant weight was depressed by K deficiency (Woltz, 1956; Lunt and Kofranek, 1958; Adams et al., 1970; Holcomb and White, 1974).

Potassium deficiency delayed flowering (Woltz, 1956; Lunt and Kofranek, 1958; Joiner and Smith, 1962; Winsor and Hart, 1965, 1966; Adams, 1975) and reduced the number of blooms on pot plants (Penningsfeld, 1972). The total number of buds formed per spray (Winsor and Hart, 1966), the number of open buds at harvest (Adams et al., 1970) and the size of the blooms (Joiner and Smith, 1962; Winsor and Hart, 1966) increased with the level of K applied. In contrast, Joiner (1967) found that bloom size was reduced by high levels of K, except when the plants were grown with high levels of P, whilst Butters and Wadsworth (1974) found that the number of buds per spray declined as the level of K increased.

Potassium deficiency reduced the proportion of large sprays; when it was not supplied as base fertiliser, the proportion of large sprays increased with the concentration of K in the liquid feed (Adams et al., 1970). The severity of the deficiency increased with successive crop-

ping; the proportion of unmarketable sprays increased from 11% in the first crop in new peat to 58% and 97% in the second and third crops respectively (Adams et al., 1971). The vase-life of the blooms after harvesting increased with the level of K at which they were grown (Butters and Wadsworth, 1974).

Potassium content of the leaves. The K content of the leaves usually declines as the plant matures e.g. Waters 1965; Sharma and Patel, 1978, particularly during flower formation (Lunt and Kofranek, 1958; Komosa, 1981). On the other hand, the K status of the leaves has been found to increase with plant age (Joiner, 1967) until the colour of the blooms was visible (Boodley and Meyer, 1965). The K content of the leaves increased with the level of potassium applied (Woltz, 1956; Lunt and Kofranek, 1958; Joiner and Smith, 1962), except at low levels of P (Joiner, 1967), but decreased with increasing levels of applied N (Waters, 1965; Carter and Dermott, 1966; Wadsworth and Butters, 1972). In healthy plants, the lower leaves contain more K than the upper leaves whereas the reverse is true for deficient plants (Lunt and Kofranek, 1958). Decreasing the root temperature depressed the K status of the leaves, particularly between 16° and 10°C (Rosenthal et al., 1973).

The leaves of deficient plants contained 0.2–2.0% K depending on the severity of the deficiency (Lunt et al., 1964), 0.50% K (Penningsfeld, 1972), 0.71–1.15% K (Holcomb and White, 1974) and less than 2.0% (Roorda van Eysinga and Smilde, 1980). For healthy plants, the ranges found or proposed include 4.0–6.0% K (Woltz, 1956), 5.0–6.0% K (Joiner and Smith, 1962), 3.5–10.0% K (Lunt et al., 1964), 3.5–6.0% K (Waters, 1965), 2.7–4.2% K (Penningsfeld, 1972), 2.2–2.7% K (Radspinner, 1974), 3.4–4.0% K (Tsujita et al., 1975), 2.6–6.1% K (Roorda van Eysinga and Smilde, 1980) and 5.3–7.5% K depending on time of year (Bunt, 1971).

Critical levels of 2.75% and 2.15% K were suggested for the upper and lower leaves respectively (Lunt et al., 1964). Komosa (1981) reported that the critical value for the laminae of the upper leaves declined from 4.5% K during vegetative growth to 3.3% K when the buds were forming and to 1.9% K when the plants were in full bloom. The need to define the 'type of critical level cited' was stressed by Hershey and Paul (1981). Thus in their study of pot-grown chrysanthemums, visible symptoms of deficiency first appeared at 0.6–0.7% K in the leaves whereas maximum growth (fresh weight) was just reached at 2.3% K.

The total uptake by winter-flowering plants was about one half (Butters and Wadsworth, 1974) or two thirds (Bunt, 1971) of that achieved by plants flowering in summer. Approximately one third of the total uptake was found in the blooms of healthy plants. In deficient plants, more of the K was found in the blooms, the proportion increasing with the severity of the disorder (Lunt and Kofranek, 1958).

Calcium

Symptoms and growth responses. The leaves of Ca deficient plants, especially the younger ones, firstly showed a slight loss of colour. This was followed rapidly by the development of brown spots, mainly around the margins of the uppermost leaves. The spots enlarged rapidly and spread across the entire leaf surface, killing the affected leaves. The growing points died and the stems also died back (Messing and Owen, 1949). The distribution of brown spots on the upper leaves and the distortion of the leaves around the growing point were well illustrated by Roorda van Eysinga and Smilde (1980). Deficient plants became stunted due to shortening of the internodes. The buds died at an early stage of development whilst the yellow leaves remained very small. The roots were stunted and died back from the tips. The margins of the leaves became brown and curled downwards as the brown spots on them coalesced. The oldest leaves remained a dull, dark green colour, were stiff and brittle, and curled downwards and inwards (Messing and Owen, 1954). Under conditions of mild deficiency the plants became stunted but the colour of the leaves was not affected. As the deficiency progressed, dark brown spots developed on the upper and lower leaves, those on the upper leaves becoming very enlarged. The blooms were very susceptible to heat injury and the stem tissue at the base of the blooms frequently collapsed (Lunt *et al.*, 1964). Woltz (1960) reported that the leaves formed a rosette and that those around the dead growing point were small, curled and thickened. The peduncles tended to collapse as the buds began to show colour. Poor keeping quality of the blooms and brown discolouration of the roots were also noted. Roorda van Eysinga and Smilde (1980) recorded that the flowers were small; the petals were poorly developed and sometimes showed brown discolouration.

Calcium deficiency reduced the height and weight of the plants, and the number of flowers formed (Penningsfeld, 1972). Liming the soil had no effect on growth or flowering (Waters, 1965), though heavy liming (to pH 7.3) reduced plant height slightly (Winsor and Hart, 1966). Most of the responses to liming involve interactions with other nutrients, as discussed elsewhere under the appropriate headings.

Calcium content of the leaves. Published data for the changes in Ca level of chrysanthemum leaves during cropping are somewhat inconsistent. Boodley and Meyer (1965) found slight increases during the first two weeks from planting, after which the levels remained fairly constant throughout two crops (spring and winter) but showed some decline followed by recovery in the middle period of a summer crop. The levels appeared to be fully adequate throughout, the values (interpolated from the author's graph) being within the range 0.8–1.5% Ca. Joiner (1967) showed an increase from an average of 0.46% Ca in the leaves two weeks after the initiation of short days to 0.84% at the time of harvest, the data being from a large factorial experiment (3 P × 3 K × 4 Mg levels) in gravel culture. Joiner and Poole (1967), however, found a very steep decline from an initial 0.78% Ca to values below 0.2% Ca at the time of flower induction, thereafter rising only slightly; the plants were grown in a 3:1 mixture of sandy soil and peat. Somewhat surprisingly, no deficiency symptoms were observed, even at a level of 0.14% Ca. The Ca content of the leaves was depressed by high levels of K (Joiner, 1967; Holcomb and White, 1974), particularly at low levels of applied N (Joiner and Smith, 1962). Calcium contents were also reduced by high levels of P or Mg (Joiner, 1967), and by low (50 mg l^{-1}) or very high (400 mg l^{-1}) levels of N in sand culture (Joiner and Smith, 1962). In solution culture, omission of N resulted in low levels of Ca in the leaves (0.31–0.37%), compared with 1.1% Ca in the 'standard' plants; foliar applications of N as urea did not compensate for lack of N supply to the roots (Meyer and Boodley, 1964). The Ca content of young chrysanthemum shoots increased as the root temperature was reduced below 16°C (Rosenthal *et al.*, 1973); the values found ranged from 0.65% Ca at ambient temperatures (18–23°C) to 0.90% Ca at 4.5°C.

The leaves of deficient plants contained 0.16–0.40% Ca (upper leaves, Lunt *et al.*, 1964), 0.2% Ca (Penningsfeld, 1972), and 1.0% Ca (Roorda van Eysinga and Smilde, 1980). For comparison, reported or recommended values for healthy plants include 0.88–3.67% Ca (Joiner and Smith, 1962), 0.50–4.60% Ca (Lunt *et al.*, 1964), 0.8–1.5% Ca (Boodley and Meyer, 1965), 0.30–1.54% Ca (Joiner, 1967), 1.49–2.53% Ca (Holcomb and White, 1974) and 1.0–2.72% Ca (Roorda van Eysinga and Smilde, 1980). Scorching on the margins of the leaves of stock plants was associated with a low status; affected leaves contained 0.30% Ca compared with 0.63% Ca in healthy leaves (Adatia, 1979). Critical values of 0.40% and 0.46% Ca respectively were deduced for upper and lower leaves (Lunt *et al.*, 1964), and a critical value of 0.5% Ca was quoted by Criley and Carlson (1970) from a commercial source.

Magnesium

Symptoms and growth responses. Although the symptoms of Mg deficiency may vary somewhat with cultivar, a sudden and rapid development of them is characteristic. The foliage first becomes pale green in colour, and then yellowing develops on the middle and lower leaves, usually interveinally, but sometimes generally. In some cultivars, small brown spots develop interveinally on the middle and lower leaves. The development of purplish-red tints on the leaves also depends on cultivar. With very young plants the foliar symptoms are severe, growth ceases, the internodes are shortened and no flowers are produced (Messing and Owen, 1954). The apex of the leaf retained its dark green colour longer than any other part of the leaf (Messing and Owen, 1949). Lunt *et al.* (1964) reported that a severe deficiency resulted in yellowing of the upper leaves; the interveinal areas later became almost white although the veins remained green. When the deficiency developed more slowly, however, the yellow margin which formed on the lower leaves gradually spread interveinally and the symptoms moved up the plant rapidly. The leaves tended to curl downwards and inwards but the roots were healthy. Roorda van Eysinga and Smilde (1980) noted interveinal chlorosis as typical for older leaves in chronic deficiency and for younger leaves in conditions of acute deficiency. Symptoms of this deficiency are less common in chrysanthemums than in other major greenhouse crops. For instance, plants grown without added Mg did not develop deficiency symptoms whereas tomatoes grown in the same plots of soil showed severe deficiency symptoms together with a depression in yield (Winsor, 1968). Lack of response to omission of Mg from the substrate was also reported by Penningsfeld (1972).

A few crop responses to Mg have been reported. For example, Mg deficiency reduced plant height (Joiner, 1967) and added Mg slightly increased the fresh weight of the leaves and the bloom diameter in soil-grown crops (Winsor, 1968). Experiments in a peat:sand mixture showed a significant reduction in stem length (Adams *et al.*, 1971), particularly at low levels of K (Adams *et al.*, 1970). Flowering was not delayed by this deficiency (Lunt *et al.*, 1964), but bloom size was reduced (Messing and Owen, 1954), though only by a severe deficiency (Lunt *et al.*, 1964). The colour of the blooms may not be affected by the deficiency (Lunt *et al.*[1], 1964), but paler colours were found with some cultivars (Messing and Owen, 1954).

Magnesium content of the leaves. Boodley and Meyer (1965) found only relatively small changes in the Mg content of chrysanthemum leaves during cropping in sand culture. Slightly lower values were recorded for a summer crop than for crops grown in winter and spring. Joiner (1967) showed marked increases in content during cropping in gravel culture e.g. from 0.64% Mg to 0.90% Mg when grown at 200 mg l^{-1} Mg in the solution. At a lower concentration (5 mg l^{-1} Mg), however, there was little change during cropping, the corresponding leaf analyses being 0.20 and 0.22% Mg respectively. Under conditions of deficiency in a soil-grown crop the leaf content decreased from 0.10% Mg eight weeks before blooming to 0.04% Mg, 7–10 days before blooming (Branson and Sciaroni, 1967).

The content of the leaves increased with the level of Mg applied (Joiner and Smith, 1962; Holcomb and White, 1974) but was depressed by very low levels of N (Joiner and Smith, 1962; Meyer and Boodley, 1964), by high levels of N (Joiner and Smith, 1962), by high levels of P (Joiner, 1967) and, particularly, by increasing levels of K (Joiner and Smith, 1962; Joiner, 1967; Holcomb and White, 1974). Root temperatures as low as 10°C did not affect the Mg content of the leaves compared with those grown at ambient root temperatures (18–23°C; Rosenthal *et al.*, 1973).

The Mg contents of the leaves decline to very low levels compared to those found in most other glasshouse crops before the plants show any symptoms of deficiency. For example, eight weeks before flowering Branson and Sciaroni (1967) found 0.1% Mg in the leaves of plants without symptoms. Five weeks later, mild deficiency symptoms had developed and the leaves contained 0.06% Mg; the symptoms became very severe as the content declined to 0.04% Mg. Deficient leaves were found to contain 0.013–0.045% Mg (lower leaves; Lunt *et al.*, 1964) and 0.15% Mg (Roorda van Eysinga and Smilde, 1980). Lunt *et al.* (1964) recorded examples of severe deficiency at 0.018% Mg and 0.030% Mg, moderate deficiency at 0.040% Mg and 0.073% Mg, slight deficiency at 0.083% Mg in the upper leaves; with one exception, the corresponding values were slightly higher in the lower leaves. For comparison, recorded or recommended values for Mg in chrysanthemum leaves include 0.3–0.9% Mg (Woltz, 1956) with a preferred level of 0.3–0.4% Mg, 0.33–1.12% Mg (Joiner and Smith, 1962), 0.06–1.50% Mg (Lunt *et al.*, 1964), 0.40–0.75% Mg (Boodley and Meyer, 1965), 0.20–0.90% Mg (Joiner, 1967), 0.39–0.67% Mg (Holcomb and White, 1974) and 0.34–1.00% Mg (Roorda van Eysinga and Smilde, 1980). Relatively high levels (0.99–1.05% Mg) were found in the leaves of plants grown at low levels of K, as for example <1.12% Mg (Joiner and Smith, 1962) and <1.05% Mg (Holcomb and White, 1974). Values of up to 1.28, 1.36 and 1.47% Mg were recorded by Lunt *et al.* (1964) under conditions of Ca deficiency.

A critical level of 0.055% Mg was suggested by Lunt *et al.* (1964), whilst Criley and Carlson (1970) quoted a value of 0.14% Mg, based on commercial experience. No symptoms were found on two cultivars of spray chrysanthemums when the leaves contained 0.12% and 0.13% Mg (Adams, unpublished data).

Iron

Symptoms and growth responses. Very mild symptoms of Fe deficiency, which resulted in some yellowing of the younger leaves, were observed by Messing and Owen (1949); growth and flowering were not affected. Woltz (1960), who noted that this deficiency was not uncommon, described interveinal yellowing of the upper leaves that gradually spread over the entire surface of severely affected leaves. Lunt *et al.* (1964) observed that the growth rate of deficient plants was very slow. As the interveinal yellowing of the younger leaves became more severe, the colour became paler until the affected leaves were almost white; large, irregularly shaped brown areas developed on these leaves. Buds formed normally, but the plants did not flower. Smilde (1975) found severe yellowing of the middle and upper leaves of plants grown in peat without added Fe, but the response varied somewhat with cultivar. However, symptoms do not always occur in peat when Fe is omitted from the fertiliser dressings (Smilde, 1972), and growth and flowering may not be affected (Adams *et al.*, 1972). When yellowing of the upper leaves of young plants is not severe, it becomes less obvious or may disappear in later stages of growth (Adams *et al.*, 1975a). Severe leaf chlorosis was illustrated by Roorda van Eysinga and Smilde (1980), who also noted that the flowers were small and pale in colour.

The responses of pot chrysanthemums to Fe deficiency varied from a delay in flowering of a few days to failure of the flower buds to develop (Bunt, 1973). Omission of Fe from the peat substrate had little effect on dry weight but delayed the flowering of both pot and spray plants. The size of the blooms on the sprays was reduced, but the number of blooms per spray and the vase-life after cutting were not affected (Smilde, 1975). The dry weight of the plants was reduced by heavy applications of Fe chelate (FeEDDHA; Smilde, 1975), and a red deposit was noticed in the lower leaves of young plants, particularly in the petioles (Adams and Winsor, unpublished data).

The response to Fe level varies with the Cu status of the plants. At adequate levels of Cu there was little response to Fe by spray chrysanthemums grown in pots of peat. However, when Cu was omitted from the peat substrate the number of axillary shoots and buds increased with the level of Fe chelate applied (Adams *et al.*, 1973). Similarly, the highest Fe level increased the dry matter content of the leaves when Cu was omitted from the fertiliser dressings, but there was no response when Cu was supplied (Davies *et al.*, 1978). At an intermediate level of Cu, applications of Fe chelate to beds of peat reduced the number of buds and flowers per spray on plants with a mild deficiency of Cu; in one instance, the proportion of marketable sprays was depressed from 76% to 4% (Adams *et al.*, 1973).

Iron content of the leaves. The Fe content of the leaves was generally decreased by liming (Smilde, 1975; Adams *et al.*, 1975b) and was greatly reduced when Fe was omitted from the nutrient solution (Lunt *et al.*, 1964). The Fe content was also increased by a very high concentration of Cu (6.4 mg l^{-1}) in solution culture (Patel *et al.*, 1976) and by omission of Cu from a peat-sand substrate (Davies *et al.*, 1978).

Healthy leaves were found to contain 100–173 μg g^{-1} Fe (Lunt *et al.*, 1964), 154–432 μg g^{-1} Fe (Smilde, 1975), 97–130 μg g^{-1} Fe (Adams *et al.*, 1975b), 65–80 μg g^{-1} Fe (Patel *et al.*, 1976) and 164–181 μg g^{-1} Fe (Davies *et al.*, 1978). No adverse response was reported when the leaves contained 518–789 μg g^{-1} Fe (Smilde, 1975).

Iron deficient leaves do not necessarily contain less Fe than healthy ones. For example, Lunt *et al.* (1964) found

35–83 μg g^{-1} Fe in completely yellow leaves whilst Roorda van Eysinga and Smilde (1980) found that the leaves became yellow when they contained less than 140 μg g^{-1} Fe. Nelson (1969) quoted a 'commonly accepted critical level of 40 ppm (40 μg g^{-1}) Fe' but found 94 μg g^{-1} Fe in the leaves of a susceptible cultivar ('Giant Betsy Rose') which showed symptoms of deficiency when grown in an alkaline vermiculite. The incidence and severity of deficiency symptoms are thus not well correlated with the total Fe content of the leaves (Bunt, 1973; Smilde, 1975), and enzymic tests for Fe may be more appropriate. Davies (1973) showed that deficiency reduced the activities of phenolase, peroxidase and indol-3-ylacetic acid (IAA) oxidase in chrysanthemum leaves. Furthermore, Davies *et al.* (1978) found that the activities of phenolase and IAA oxidase both increased markedly with the level of Fe applied. Peroxidase activity was increased by an intermediate level of added Fe but did not respond further to a higher rate of applicaion.

Manganese

Symptoms and growth responses. Plants with Mn deficiency made less vigorous growth and had weaker stems and paler green leaves than healthy plants. The symptoms which developed later on the middle and upper leaves varied somewhat with cultivar but usually took the form of interveinal chlorosis. As the veins, including the smaller ones, remained green, the leaves had a characteristic 'netted' appearance. The margins and tips of the middle leaves turned downwards in a characteristic manner. Reddish-brown areas developed within the chlorotic zones and soon dried out (Messing and Owen, 1951). The colour and texture of the blooms were adversely affected; the petals lacked crispness and the flowers did not keep well (Messing and Owen, 1954). Woltz (1960) reported mild, interveinal chlorosis on the young leaves. Small interveinal spots developed on the middle leaves, these at first being white or gray and later tan coloured. Lunt *et al.* (1964) found that foliage formed under deficiency conditions was a pale lime-green colour with no veinal pattern; the leaves were small and the growth spindly. White spots appeared on the upper leaves, these being most pronounced near the margins. Larger, bronzed patches, which gradually dried out, developed at random on the older leaves. The root system was healthy but reduced in size. Adams *et al.* (1975a) noted that the middle and lower leaves of some spray cultivars showed symptoms of the deficiency, namely a netted appearance due to interveinal yellowing and down-curling of the margins. Roorda van Eysinga and Smilde (1980) referred to a general loss of plant colour, the leaves turning uniformly pale green, but also included a second photograph showing interveinal chlorosis.

Manganese deficiency delayed flowering in sand or solution culture (Messing and Owen, 1954; Lunt *et al.*, 1964) and reduced the size of the blooms (Messing and

Owen, 1954). The omission of Mn from an otherwise completely fertilised peat had no effect on growth or flowering (Smilde, 1972; Adams *et al.*, 1972; Davies *et al.*, 1978), but when added it reduced the number of buds developing on sprays with a low Fe and Cu status (Adams *et al.*, 1973).

Manganese content of the leaves of deficient and normal plants. The Mn content of the leaves generally increased with the age of the plants (Sharma and Patel, 1978) and with the level of application (Davies *et al.*, 1978). Manganese content was depressed by liming (Smilde, 1975; Adams *et al.*, 1975b) and by dressings of Fe and Cu (Davies *et al.*, 1978). Manganese deficiency reduced the activity of phenolase in the leaves (Davies, 1973), but had no effect on the activities of peroxidase and IAA oxidase (Davies *et al.*, 1978).

The leaves of deficient plants contained only 3–4 μg g^{-1} Mn (Lunt *et al.*, 1964) and 15–20 μg g^{-1} Mn (Roorda van Eysinga and Smilde, 1980). For comparison, Lunt *et al.* (1964) recorded 195 and 260 μg g^{-1} Mn respectively in the upper and lower leaves of healthy plants, these receiving the standard treatment (0.5 mg l^{-1} Mn) in their solution culture studies. Values of 146–283 μg g^{-1} Mn were found by Adams *et al.* (1975b) in the leaves of spray chrysanthemums grown with two levels of a fritted trace element mixture at pH values of 4.2–5.1 in the peat:sand substrate, decreasing to 34–40 μg g^{-1} at pH 6.0 and to 29 30 μg g^{-1} at pH 6.9. Plants grown without added trace elements at high pH (6.4–7.2) contained only 21–23 μg g^{-1} Mn. Patel *et al.* (1976) reported 61 μg g^{-1} Mn in the leaves of the young 'standard' plants grown with 0.25 mg l^{-1} Mn in solution culture. The corresponding values for stems and roots were 4.7 and 129 μg g^{-1} Mn. Davies *et al.* (1978) found values of 67–482 μg g^{-1} Mn in the leaves of chrysanthemums receiving various combinations of Mn, Fe and Cu. There was no indication of deficiency, though the high levels of Mn were unfavourable to flower production when Cu was also deficient. Roorda van Eysinga and Smilde (1980) recorded a range of 15–250 μg g^{-1} Mn in healthy chrysanthemum leaves.

Sharma and Patel (1978) referred to a (lower) critical level of 200 μg g^{-1} Mn in chrysanthemum leaves, based on data quoted by Criley and Carlson (1970) from a commercial source. On this basis, Sharma and Patel regarded some of their analytical data, which covered the range 47–565 μg g^{-1} Mn, as indicating a deficiency. A critical level of 200 μg g^{-1} Mn would, however, appear to be unduly high (Adams *et al.*, 1975b).

Manganese toxicity. Woltz (1956) induced toxicity in chrysanthemums grown in sand culture. High concentrations of Mn in solution (8–16 mg l^{-1}) produced a chlorotic condition, particularly in the younger leaves, described as somewhat similar to Fe deficiency. Manganese toxicity frequently occurs after soils have been steam-sterilised.

Winsor and Hart (1965) reported yellowing of the young leaves soon after planting into slightly acid soil; the symptoms persisted throughout cropping. The incidence of symptoms increased with the level of N, presumably due to the acidification associated with ammonium nitrate. In one cultivar, small brown spots developed on all the leaves, including the smallest leaflets in the head, but were most widespread on mature leaves. Roorda van Eysinga and Smilde (1980) noted that growth was restricted and that the yellowing on the upper leaves resembled that due to Fe deficiency, the main veins remaining green at first. The small brown spots which developed on the older leaves spread inwards from the margins until the entire leaf surface was desiccated. Fewer buds were formed on the affected plants and the blooms remained small.

Winsor (1972) noted that Mn toxicity was not obviated by liming to pH 7.3 when the soil at the bottom of raised concrete beds was poorly aerated, due to excessive moisture levels. Adatia (1979) found that yellowing of the lower leaves of pot chrysanthemums was due to the very moist conditions maintained by capillary watering, despite a soil pH of 7.1. The affected leaves developed brown spots on the laminae with blackish deposits in the veins. Tsujita *et al.* (1975) found that the Mn content of the leaves was doubled when ammonium nitrate was used instead of calcium nitrate as the N source.

The yellow leaves of plants affected by the toxicity contained up to 2600 μg g^{-1} Mn (Winsor and Hart, 1965) and 800–980 μg g^{-1} Mn (Adatia, 1979). Concentrations of 852 μg g^{-1} Mn were associated with toxicity in some cultivars, though others tolerate much higher concentrations (Roorda van Eysinga and Smilde, 1980). Woltz (1956) recorded values up to 2600 μg g^{-1} Mn in chlorotic plants, and Taylor and Joiner (1963) found 4155 μg g^{-1} Mn in the leaves of plants receiving 50 mg l^{-1} Mn in sand culture. Chrysanthemums are fairly tolerant to high levels; thus Bunt (1972) and Sonneveld and Voogt (1975) found that leaves containing 900–1000 μg g^{-1} Mn did not show symptoms of the toxicity, and Smilde (1975) found no evidence of toxicity at concentrations of up to 1500 μg g^{-1} Mn.

Copper

Symptoms and growth responses. Recent studies of responses to Cu have been more extensive than those with other micronutrients, due to the involvement of this nutrient in the flowering process (see, for example, Davies, 1973; Davies *et al.*, 1978).

Copper deficiency in chrysanthemums was described by Woltz (1960). Deficient plants had dull green leaves. Yellowing of the veins developed and spread outwards to the margins, and the affected leaves wilted during the day. The flowers were small and lacked crispness. With a mild deficiency the plants were slightly spindly, had longer internodes, longer leaves, were taller and flowered 4–5 days later than normal plants (Lunt *et al.*, 1964). Yellowing near the margins, which included the fine veins, occurred first on the middle leaves, but progressed slowly to the younger leaves during flower bud development. The first two leaves on basal shoots became yellow and the shoots made no further growth whereas those higher up grew well. Penningsfeld and Forchthammer (1970) illustrated a serious delay in budding on plants grown without added Cu. The middle and older leaves had withered away, leaving only the upper leaves and those on the original cutting. Nelson (1971) reported that the young leaves were paler than normal. Interveinal yellowing developed, except at the leaf margins and on the lobes near the leaf tip. When the deficiency became severe, the tissue near the veins became yellow and growth was stunted. Brown areas developed on the mature leaves and, except for the leaves on the original cutting, those below the first expanded leaf all died. The flowers had an untidy appearance and were small, with fewer petals than normal. Smilde (1971, 1975) reported that a deficiency caused yellowing and desiccation of the middle leaves and delayed the flowering of pot chrysanthemums.

Adams et al. (1971) observed that the normal darkening of the foliage associated with flower bud formation did not occur on deficient plants; the leaves remained pale green and budding was either delayed or prevented. No visible buds were formed on plants stunted by severe deficiency, and very few axillary shoots were produced (Adams et al., 1972). Graves and Sutcliffe (1974) dissected the growing points of deficient and healthy plants of the same age in order to assess bud development. They found that deficiency delayed bud initiation and that there was little subsequent development of buds that had been seriously delayed.

The descriptions reported so far indicate a considerable variation in the symptoms of this deficiency which is partly due to large differences in the symptoms of mild and severe forms. With the milder form, which has been described and illustrated in some detail (Adams, 1976), there is a tendency for the axillary shoots to branch excessively during the delay in budding. They thus become larger and bear more buds than healthy plants, but are of little decorative value. Adams et al. (1975c; see Table 26) found that, with a severe deficiency, the axillary shoots produced numerous bracts and leaves as compared with the one or two small bracts below the flowers on healthy plants. In this respect Cu deficiency simulates the effects of continuous long days, but differs by causing collapse and death of some of the middle and lower leaves on the main stem. With a more severe deficiency, few axillary shoots develop; these bear pale green but otherwise normal leaves characteristic of the cultivar, as opposed to the enlarged bracts formed when budding is seriously delayed (illustrated; Adams et al., 1975c). Bud formation is suppressed, and the middle and lower leaves on the main stem wither. With a very severe deficiency, all the leaves die excepting a few immediately surrounding the

growing point and those on the original cutting. In extreme cases the growing point itself dies (Adams, 1975). Some variation in the response of different cultivars to the deficiency has been reported for plants grown as sprays, standards and pot chrysanthemums (Adams et al., 1975; Adams et al., 1976; Graves et al., 1979).

Copper deficiency caused a marked reduction in the dry matter content of the leaves (Adams et al., 1975c; Davies et al., 1978). The effects of the deficiency on flowering ranged from a few days delay with pot plants (10 days, Bunt, 1973; 4–10 days, Graves et al., 1979) to complete suppression of flowering (Adams et al., 1972, 1975c). The less serious effect of the deficiency on the pot plants appears to be linked with the normal commercial practice of using a growth retardant. When Cu was not supplied the untreated plants were appreciably taller and budding was more seriously delayed than on plants treated with a retardant (Adams et al., 1975). With spray chrysanthemums, only 7–15% of the plants in deficiency treatments had flowered some 25 days after all plants receiving Cu had been harvested (Adams et al., 1975c). This response was simulated by applying IAA, a growth hormone, to the upper leaves of chrysanthemums; bud initiation on treated plants was delayed by up to 40 days (Graves et al., 1977). This finding lends support to a hypothesis proposed by Davies and co-workers (Davies, 1973; Adams et al., 1975c; Davies et al., 1978) to explain the effect of Cu on the flowering of chrysanthemums. Deficiency limits the synthesis of phenolase, a Cu containing enzyme. Phenolic compounds thus accumulate in the tissue and these in turn prevent the decomposition of IAA by decreasing the activity of IAA-oxidase. In support of this concept it was found that Cu deficiency depressed phenolase activity in the leaves by almost 80%, IAA oxidase activity by 55% and peroxidase activity by 36%. Unfortunately, attempts to measure the concentration of IAA in healthy and deficient chrysanthemum tissue were unsuccessful. The relationship between phenolase activity and time of flowering was supported by the work of Graves et al. (1979). Both normal and deficient plants treated with a growth retardant had a higher phenolase activity than the corresponding untreated plants; greater delay in the flowering of plants not treated with the growth regulator was associated with lower phenolase activity.

Flowering ability was restored to deficient plants when Cu was subsequently supplied. However, the number of blooms eventually produced per spray was fewer than normal, depending on the severity of the deficiency and the length of time that elapsed before it was supplied (Graves et al., 1978).

There was little direct response to heavy applications of Cu (Adams et al., 1977); severe toxicity had been reported elsewhere (Patel et al., 1976) but not described. An interaction of excess Cu and B level has been reported for one cultivar (Adams et al., 1977); a high level of Cu reduced the delay in flowering due to B deficiency.

Treatment	Stem length at harvest (cm)	Leaves on main stem	Length of second shoot from top (cm)	Leaves on second shoot from top	Flowering shoots per spray	Large sprays (%)	Rejects (%)	Copper content (μg g^{-1})
Complete micronutrients	65	29	5.3	0	9.3	84	2	8.0
− B	66	29	5.5	0	8.6	68	3	7.4
− Cu	76	36	20.9	10	0.0	1	86	2.0
− B, − Cu	75	39	20.5	10	0.1	1	93	1.9
Nil	67	34	20.0	8	1.0	0	90	2.5

Table 26. Some effects of micronutrient treatments on the growth, flowering and copper content of spray chrysanthemums (after Adams, Graves and Winsor, 1975c).

Copper content of the leaves. The Cu content of the leaves declined progressively as the root temperature was reduced from ambient (18–23°C) to 4.5°C (Rosenthal *et al.*, 1973). The content of the leaves of healthy plants generally declined with age and was negatively correlated with the amount of dry matter produced (Graves, 1978). The content was therefore highest in young leaves near the growing point and lower in the older leaves. The reverse was found to be true for Cu deficient plants. It was suggested that the critical level with respect to flowering was the content of the young, rapidly expanding leaves rather than the Cu status of the whole plant.

During vegetative growth, leaves with deficiency symptoms contained slightly less than 5 μg g^{-1} Cu whereas the upper leaves contained 8.5–10.0 μg g^{-1} Cu; with severe deficiency the middle leaves contained 1.7–4.7 μg g^{-1} Cu (Lunt *et al.*, 1964). Other published values include 2.9 μg g^{-1} Cu (severe deficiency; Penningsfeld and Forchthammer, 1970), 3.5 μg g^{-1} Cu (Adams *et al.*, 1972), 1–4 μg g^{-1} Cu (Adams *et al.*, 1974a), 1.8–2.4 μg g^{-1} Cu (Adams *et al.*, 1975b), 1.4–2.5 μg g^{-1} Cu (Adams *et al.*, 1975c), 4.5–5.0 μg g^{-1} Cu (Smilde, 1975), 2.3–2.6 μg g^{-1} Cu (Davies *et al.*, 1978) and 1–4 μg g^{-1} Cu (Graves *et al.*, 1979). Somewhat higher values were found for leaves from deficient pot plants, for example 6.4 μg g^{-1} Cu (Bunt, 1973) and 3–11 μg g^{-1} Cu (Graves *et al.*, 1979).

The leaves of healthy plants contained 10 μg g^{-1} Cu (Lunt *et al.*, 1964), 20.5 μg g^{-1} Cu (pot plants; Bunt, 1973), 7–20 μg g^{-1} Cu (Adams *et al.*, 1974a), 5–14 μg g^{-1} Cu (Adams *et al.*, 1975b), 7–14 μg g^{-1} Cu (Adams *et al.*, 1975c), 6–10 μg g^{-1} Cu (Smilde, 1975), 7–12 μg g^{-1} Cu (Patel *et al.*, 1976), 7.4–8.2 μg g^{-1} Cu (Davies *et al.*, 1978) and 7–19 μg g^{-1} Cu (Graves *et al.*, 1979).

Critical values of 5 μg g^{-1} Cu (Lunt *et al.*, 1964), 6.7–7.4 (Nelson, 1971) and 4 μg g^{-1} Cu (Graves *et al.*, 1979) have been suggested. Severe toxicity was found when the leaves contained more than 48 μg g^{-1} Cu (Patel *et al.*, 1976).

Zinc

Symptoms and growth responses. Symptoms of Zn deficiency developed, shortly before flowering, in the form of small yellow spots on the middle and upper leaves.

These spots enlarged, became brown in the centre and spread progressively to younger leaves at flowering (Lunt *et al.*, 1964). Growth became stunted, giving the plants a rosette appearance. The middle and lower leaves curled downwards and the yellow spots coalesced to give an irregular pattern. When the deficiency was severe, brown spots developed in the yellow areas. The flowers remained small (Roorda van Eysinga and Smilde, 1980). These symptoms were produced in sand or solution culture. Omission of Zn from a peat substrate had no apparent effect on growth or flowering of sprays (Adams *et al.*, 1972), but reduced the growth and bloom size of pot chrysanthemums (Penningsfeld, 1972).

Zinc content of the leaves. The content of the leaves was appreciably higher (68%) when ammonium nitrate was used instead of calcium nitrate (Tsujita *et al.*, 1975). Reducing the temperature of the roots caused a progressive reduction in the content of the leaves, the values reported ranging from 86 μg g^{-1} at ambient temperatures (18–23°C) to 37 μg g^{-1} Zn at 4°C (Rosenthal *et al.*, 1973).

The leaves of deficient plants contained 4.3–7.8 μg g^{-1} Zn (lower leaves; Lunt *et al.*, 1964) and 7.2 μg g^{-1} Zn (Roorda van Eysinga and Smilde, 1980) compared with 7–26 μg g^{-1} Zn (Lunt *et al.*, 1964), 54–67 μg g^{-1} Zn (Penningsfeld, 1972), 135–472 μg g^{-1} Zn (Smilde, 1975), 43–110 μg g^{-1} Zn (Adams *et al.*, 1975c), 37–75 μg g^{-1} Zn (Patel *et al.*, 1976) and 92 μg g^{-1} Zn (Roorda van Eysinga and Smilde, 1980) in healthy leaves. No adverse effects were reported when the leaves contained 319–880 μg g^{-1} Zn (Smilde, 1975). A critical value of 7 μg g^{-1} was given by Lunt *et al.* (1964).

Symptoms of excess Zn, described and illustrated by Roorda van Eysinga and Smilde (1980), include restricted growth, interveinal yellowing and irregular brown spots at the margins of the older leaves. The plants tended to wilt easily due to the unhealthy root system. The symptoms of toxicity on young plants are similar to those of Fe deficiency, the youngest leaves being completely yellow.

Growth may be affected when the leaves contain more than 900 μg g^{-1} Zn (Smilde, 1975). Patel *et al.* (1976) recorded a 20% reduction in the growth of young plants when the level in the leaves was raised to 421 μg g^{-1}; the corresponding values for stems and roots were 375 and 500 μg g^{-1} Zn. Distinct symptoms of the toxicity were found on leaves containing 1458 μg g^{-1} Zn (Roorda van Eysinga and Smilde, 1980).

Boron

Symptoms and growth responses. Detailed descriptions of B deficiency were given by Messing and Owen (1949, 1951, 1954). Boron deficient plants had short internodes and pale upper leaves, with interveinal yellowing. Down-curling and inrolling of the leaves, together with brittle-ness, were characteristic foliar symptoms, and purple-red tints developed on some cultivars. Browning and cracking of the veins were also observed, and the roots were short and tended to die back from the tips. In cases of severe deficiency, 40% of the shoots failed to produce blooms and, with an extreme deficiency, the growing points died (Messing and Owen, 1954). Woltz (1960) recorded inter-veinal yellowing of the leaves, with anthocyanins in the veins. It was noted that browning of the roots occurred before the foliar symptoms appeared. Lunt et al. (1964) found that the middle leaves, and later the upper leaves, became slightly yellow and developed a smooth leathery texture. Reddish spots developed on the stem near the leaf axils. Graves (1971) found that a severe deficiency shortened the plants by reducing both the number and the length of the internodes on the stem. The lower leaves became yellow with purple tints, the veins were brown, and the roots were stunted and tended to die back from the tips. Boron deficiency had little effect on the foliage of pot plants (Bunt, 1965). The flower stems were extremely brittle, however, and the terminal bud tended to break off when the axillary buds were removed. Under conditions of severe deficiency the buds failed to open (Messing and Owen, 1949) or to open properly (Adams, 1975).

The petals on deficient plants were curled upwards and inwards ('quilled'; Messing and Owen, 1954; Bunt, 1965; Adams et al., 1975b; see Figure 26). Certain 'standard' varieties which normally had reflexed blooms became incurved, and those which were normally incurved became more tightly packed (Messing and Owen, 1954; see also Woltz, 1960; Bunt, 1965). The surface of the petals was dull and rough (Messing and Owen, 1954; Woltz, 1960) and the petals bruised easily and became brown (Messing and Owen, 1954; Woltz, 1960; Boodley and Sheldrake, 1973). Blackening, which was not associated with fungal or bacterial infection, developed in the centre of some blooms (Bunt, 1965).

Boron deficiency delayed flowering (Graves, 1971; Smilde, 1975; Adams et al., 1975c) and reduced the size of the blooms (Messing and Owen, 1954), the number of flowering shoots per spray (Graves, 1971) and the propor-tion of large sprays (see Table 26). Omission of both B and Cu together from the peat substrate accentuated the effects of Cu deficiency (Adams et al., 1975c, 1976).

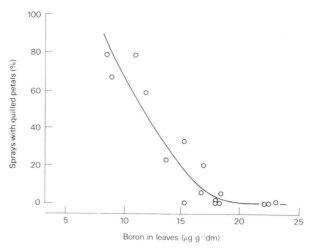

Figure 26   Relation between the boron content of the foliage and the proportion of spray chrysanthemums with quilled petals when grown in peat without added micronutrients (after Adams et al., 1975a).

Boron content of the leaves.   Deficient leaves contained 18–20 μg g^{-1} B (Lunt et al., 1964), 4.6–6.2 μg g^{-1} B (severe deficiency; Graves, 1971), 15 μg g^{-1} B (Penning-sfeld, 1972; Smilde, 1975), 8–20 μg g^{-1} B (Adams et al., 1975b; see Fig. 26), 3–13 μg g^{-1} B (Adams et al., 1975c) and less than 25 μg g^{-1} B (Roorda van Eysinga and Smilde, 1980). For comparison, healthy leaves contained 75 μg g^{-1} B (Woltz, 1956), 25–200 μg g^{-1} B (Lunt et al., 1964), 25–30 μg g^{-1} B (Smilde, 1975), more than 20 μg g^{-1} B (Adams et al., 1975b), 35–63 μg g^{-1} (Adams et al., 1975c) and 62 μg g^{-1} B (Roorda van Eysinga and Smilde, 1980). The content of chrysanthemum leaves was de-creased progressively by liming (Adams et al., 1975b; see Fig. 27).

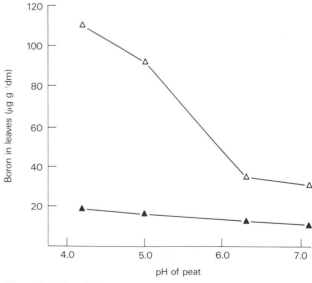

Figure 27   Effect of pH on the boron content of the leaves of chrysanthemums grown in peat with (△) and without (▲) a fritted micronutrient mixture (after Adams et al., 1975b).

Excessive levels of B result in browning of the margins, which spreads over the whole surface of the lower leaves; the symptoms spread up the plant and, at very high levels, cause injury to the buds (Woltz, 1956). Lunt *et al.* (1964) reported that black spots developed and coalesced along the margins of the lower leaves, causing the margins to curl downwards. Gogue and Sanderson (1973) found that reddish-brown interveinal areas developed near the margins of the lower leaves and spread inwards towards the mid-rib; interveinal yellowing developed on the upper leaves. Smilde (1975) recorded necrotic spotting and curling of the leaf edges. Excellent illustrations of the marginal browning were published by Roorda van Eysinga and Smilde (1980).

Boron toxicity decreased stem length and flower diameter; growth was reduced when the leaves contained more than 100 $\mu g\,g^{-1}$ B (Gogue and Sanderson, 1973). Excess B depressed the total number of buds formed by one cultivar (Adams *et al.*, 1977, 1978) and delayed flowering, the latter response being accentuated by high levels of Cu (Adams *et al.*, 1977). Leaves with toxicity symptoms contained 380 $\mu g\,g^{-1}$ B (Woltz, 1956), 240 $\mu g\,g^{-1}$ B or more (Lunt *et al.*, 1964), 136 $\mu g\,g^{-1}$ B (Gogue and Sanderson, 1973), more than 120 $\mu g\,g^{-1}$ B (Smilde, 1975) and more than 98 $\mu g\,g^{-1}$ B (Roorda van Eysinga and Smilde, 1980).

Molybdenum

Little has been published about the responses of chrysanthemums to this nutrient. The reports available suggest that omission of Mo from the fertilisers added to a peat substrate had little effect on growth or flowering (Penningsfeld, 1972; Adams *et al.*, 1972), though a slight reduction in the dry weight was found under acid conditions (pH 4.4; Smilde, 1975).

The Mo content of the leaves increased with the level of sodium molybdate applied and with the pH of the peat substrate (Smilde, 1975). The leaves of deficient plants contained 0.05 $\mu g\,g^{-1}$ Mo compared with 0.5 $\mu g\,g^{-1}$ Mo in healthy leaves (Penningsfeld and Forchthammer, 1970). No symptoms of deficiency or toxicity were reported when the leaves contained 0.1–1.2 $\mu g\,g^{-1}$ Mo (Penningsfeld, 1972) or 0.1–18.7 $\mu g\,g^{-1}$ Mo (Smilde, 1975). Heavy dressings of sodium molybdate applied to a peat substrate (pH 6.9) had no harmful effects, although the leaves contained 116 $\mu g\,g^{-1}$ Mo (Smilde, 1975).

References

Adams, P. (1975). Concise descriptions of nutritional disorders. In: Pests, Diseases and Nutritional Disorders of the Chrysanthemum by Scopes, N. *Nat Chrysanthemum Soc, Lond*, 67,–76. (Reproduced in Chrysanthemums, Year-round Growing, 1978. Machin, B. and Scopes, N., *Blandford Press*).

Adams, P. (1976). Copper deficiency in chrysanthemums. In: The Chrysanthemum Year Book, 1976. *Nat Chrysanthemum Soc Lond*, 82–91. (Reprinted in The Chrysanthemum 1977, **33**, *Nat*

Chrysanthemum Soc Inc. USA, 127–133, New York.

Adams, P., Davies, J. N., Graves, C. J. and Winsor, G. W. (1974). Copper and the flowering of chrysanthemums. *J Sci Fd Agric* **25**, 1192.

Adams, P., Graves, C. J. and Winsor, G. W. (1970). Nutrition of year-round chrysanthemums in peat and sand. *Rep Glasshouse Crops Res Inst 1969*, 90–93.

Adams, P., Graves, C. J. and Winsor, G. W. (1971). Nutrition of year-round chrysanthemums in peat and sand. *Rep Glasshouse Crops Res Inst 1970*, 99–103.

Adams, P., Graves, C. J. and Winsor, G. W. (1972). Nutrition of year-round chrysanthemums in peat/sand compost. *Rep Glasshouse Crops Res Inst 1971*, 79–80.

Adams, P., Graves, C. J. and Winsor, G. W. (1973). Micronutrients affecting the flowering of chrysanthemums *Chrysanthemum morifolium* Ramat. *Rep Glasshouse Crops Res Inst 1972*, 62.

Adams, P., Graves, C. J. and Winsor, G. W. (1975a). Nutrition of year-round chrysanthemums in peat. *Rep Glasshouse Crops Res Inst 1974*, 69.

Adams, P., Graves, C. J. and Winsor, G. W. (1975b). Some responses of *Chrysanthemum morifolium* (cv. "Hurricane"), grown as a year-round crop in a peat-sand substrate, to micronutrients and liming. *J Sci Fd Agric* **26**, 769–778.

Adams, P., Graves, C. J. and Winsor, G. W. (1975c). Some effects of copper and boron deficiencies on the growth and flowering of *Chrysanthemum morifolium* (cv. Hurricane). *J Sci Fd Agric* **26**, 1899–1909.

Adams, P., Graves, C. J. and Winsor, G. W. (1976). Nutrition of year-round chrysanthemums in peat. *Rep Glasshouse Crops Res Inst 1975*, 72–73.

Adams, P., Graves, C. J. and Winsor, G. W. (1977). Nutrition of year-round chrysanthemums in peat. *Rep Glasshouse Crops Res Inst 1976*, 80.

Adams, P., Graves, C. J. and Winsor, G. W. (1978). Nutrition of year-round chrysanthemums in peat. *Rep Glasshouse Crops Res Inst 1977*, 81–83.

Adams, P. and Winsor, G. W. (1967). Chrysanthemum nutrition: responses to phosphate and lime. *Rep Glasshouse Crops Res Inst 1966*, 71–72.

Adatia, M. H. (1979). Advisory and analytical service. *Rep Glasshouse Crops Res Inst 1978*, 90–91.

Boodley, J. W. and Meyer, M. (1965). The nutrient content of Bonnaffon Deluxe chrysanthemums from juvenile to mature growth. *Proc Am Soc hort Sci* **87**, 472–478.

Boodley, J. W. and Sheldrake, Jr., R. (1973). Boron deficiency and petal necrosis of 'Indianapolis White' chrysanthemum. *HortSci* **8**, 24–26.

Branson, R. and Sciaroni, R. H. (1967). Magnesium deficiency in cut flower chrysanthemums. *Colo Flow Gro Assn Bull* **203**, 3.

Bunt, A. C. (1965). Look for boron deficiency in peat-sand composts. *Grower* **64**, 440–441.

Bunt, A. C. (1971). The use of peat-sand substrates for pot chrysanthemum culture. *Acta Hort* **18**, 66–74.

Bunt, A. C. (1972). The use of fritted trace elements in peat-sand substrates. *Acta Hort* **26**, 129–140.

Bunt, A. C. (1973). Factors contributing to the delay in the flowering of pot chrysanthemums grown in peat-sand substrates. *Acta Hort* **31**, 163–172.

Butters, R. E. and Wadsworth, G. A. (1974). Nutrition of year-round spray chrysanthemums in beds of soil-less composts. *Expl Hort* **26**, 17–31.

Carter, A. R. and Dermott, W. (1966). Nitrogen manuring of early flowering chrysanthemums. I. In unheated glass. *Expl Hort* **16**, 12–18.

Criley, R. A. and Carlson, W. H. (1970). Tissue analysis standards for various floricultural crops. *Florists' Rev* **146**, 19–20, 70–73.

Davies, J. N. (1973). Enzymes in chrysanthemum tissue. *Rep Glasshouse Crops Res Inst 1972*, 62–63.

Davies, J. N., Adams, P. and Winsor, G. W. (1978). Bud development and flowering of *Chrysanthemum morifolium* in relation to some enzyme activities and to the copper, iron and manganese status. *Commun Soil Sci Pl Anal* 9, 249–264.

Gogue, G. J. and Sanderson, K. C. (1973). Boron toxicity of chrysanthemum. *HortSci* 8, 473–475.

Graves, C. J. (1971). A study of boron deficiency in chrysanthemums grown in solution culture. *Rep Glasshouse Crops Res Inst 1970*, 107–108.

Graves, C. J. (1978). Uptake and distribution of copper in *Chrysanthemum morifolium*. *Ann Bot* 42, 117–125.

Graves, C. J. and Sutcliffe, J. F. (1974). An effect of copper deficiency on the initiation and development of flower buds of *Chrysanthemum morifolium* grown in solution culture. *Ann Bot* 38, 729–738.

Graves, C. J., Adams, P. and Winsor, G. W. (1977). Some effects of indol-3-ylacetic acid on the rates of initiation and development of flower buds of *Chrysanthemum morifolium*. *Ann Bot* 41, 747–753.

Graves, C. J., Adams, P. and Winsor, G. W. (1978). The regulation of flowering by copper in *Chrysanthemum morifolium*. *Ann Bot* 42, 1241–1243.

Graves, C. J., Adams, P. and Winsor, G. W. (1979). Some effects of copper deficiency on the flowering, copper status and phenolase activity of different cultivars of *Chrysanthemum morifolium*. *J Sci Fd Agric* 30, 751–758.

Harbaugh, B. K., Price, J. F. and Stanley, C. D. (1983). Influence of leaf nitrogen on leafminer damage and yield of spray chrysanthemum. *HortSci* 18, 880–881.

Hershey, D. R. and Paul, J. L. (1981). Critical foliar levels of potassium in pot chrysanthemum. *HortSci* 16, 220–222.

Holcomb, E. J. and White, J. W. (1974). Potassium fertilisation of chrysanthemums using a constant-drip fertiliser solution. *Pl Soil* 41, 271–278.

Joiner, J. N. (1967). Effects of P, K and Mg levels on growth, yield and chemical composition of *Chrysanthemum morifolium* 'Indianapolis White no. 3'. *Proc Am Soc hort Sci* 90, 389–396.

Joiner, J. N. and Poole, R. T. (1967). Relationship of fertilisation frequency to chrysanthemum yield and nutrient levels in soil and foliage. *Proc Am Soc hort Sci* 90, 397–402.

Joiner, J. N. and Smith, T. C. (1961). Some effects of nitrogen and potassium levels on flowering characteristics of *Chrysanthemum morifolium* 'Bluechip'. *Proc Fla St hort Soc* 73, 354–358.

Joiner, J. N. and Smith, T. C. (1962). Effects of nitrogen and potassium levels on the growth, flowering responses and foliar composition of *Chrysanthemum morifolium* 'Bluechip'. *Proc Am Soc hort Sci* 80, 571–580.

Komosa, A. (1981). Low and high critical levels of nitrogen, phosphorus and potassium for chrysanthemum *(Chrysanthemum morifolium* cv. Balcombe Perfection). *Acta Hort* 125, 61–68.

Lunt, O. R. and Kofranek, A. M. (1958). Nitrogen and potassium nutrition of chrysanthemum. *Proc Am Soc hort Sci* 72, 487–497.

Lunt, O. R., Kofranek, A. M. and Oertli, J. J. (1946). Some critical nutrient levels in *Chrysanthemum morifolium*, cultivar Good News. In: Plant analysis and fertiliser problems IV (Bould, C., Prevot, P., Magness, J. R., Eds). *Am Soc hort Sci, East Lansing 1964*, 398–413.

Machin, B. (1982). Improved nutrition can cut costs. *Grower* 97, No. 18, 22–23.

Massey, D. M. and Winsor, G. W. (1969). Nutrition of year-round chrysanthemums. *Rep Glasshouse Crops Res Inst 1968*, 71–72.

Massey, D. M. and Winsor, G. W. (1974). The nitrogen nutrition of spray chrysanthemums (cv. Hurricane) grown as a year-round crop in soil borders. *Expl Hort* 26, 1–16.

Messing, J. H. L. and Owen, O. (1949). The effect of some acute mineral deficiencies on glasshouse plants. *Rep Exp Res Stn Cheshunt 1948*, 73–80.

Messing, J. H. L. and Owen, O. (1951). The effects of some acute mineral deficiencies on glasshouse chrysanthemums. *Rep Exp Res Stn Cheshunt 1950*, 57–60.

Messing, J. H. L. and Owen, O. (1954). The visual symptoms of some mineral deficiencies on chrysanthemums. *Pl Soil* 5, 101–120.

Meyer, M. M. and Boodley, J. W. (1964). Foliar applications of nitrogen, phosphorus and potassium to chrysanthemum and poinsettia. *Proc Am Soc hort Sci* 84, 582–587.

Nelson, P. V. (1969). Assessment and correction of the alkalinity problem associated with Palabora vermiculite. *J Am Soc hort Sci* 94, 664–667.

Nelson, P. V. (1971). Copper deficiency in chrysanthemum: Critical level and symptoms. *J Am Soc hort Sci* 96, 661–663.

Nelson, P. V. and Hsieh, K. (1971). Ammonium toxicity in chrysanthemum: critical level and symptoms. *Commun Soil Sci Pl Anal* 2, 439–448.

Nichols, R. (1970). Pre-harvest nutrition and keeping quality of chrysanthemums. *Rep Glasshouse Crops Res Inst 1969*, 68.

Patel, P. M., Wallace, A. and Mueller, R. T. (1976). Some effects of copper, cobalt, cadmium, zinc, nickel and chromium on growth and mineral element concentration in chrysanthemum. *J Am Soc hort Sci* 101, 553–556.

Penningsfeld, F. (1972). Macro and micro nutrient requirements of pot plants in peat. *Acta Hort* 26, 81–101.

Penningsfeld, F. and Forchthammer, L. (1970). Essais pour la détermination de l'approvisionnement optimal en oligoéléments des fleurs à couper. *Proc 6th Colloq Int Pot Inst, Florence 1968*, 364–378.

Radspinner, A. L. (1974). Effects of levels of calcium and iron and sources of phosphorus on growth and leaf composition of *Chrysanthemum morifolium* 'Iceberg'. *CEIBA* 18, 33–40.

Roorda van Eysinga, J. P. N. L. and Smilde, K. W. (1980). Nutritional disorders in chrysanthemums. *Cent agric Publ & Docum Wageningen*, 42 pp.

Rosenthal, R. N., Woodbridge, C. G. and Pfeiffer, C. L. (1973). Root temperature and nutrient levels of chrysanthemum shoots. *HortSci* 8, 26–27.

Sharma, G. C. and Patel, A. J. (1978). Effect of nine controlled-release fertilisers on chrysanthemum growth and foliar analysis. *J Am Soc hort Sci* 103, 148–150.

Smilde, K. W. (1971). Evaluation of fritted trace elements (fte) on peat substrates. Rapport 1–1971, *Instituut voor bodemvruchtbaarheid, Haren – Groningen*, 15 pp.

Smilde, K. W. (1972). Trace nutrient requirements of some plant species on peat substrates. *Proc 4th Int Peat Cong Helsinki 1972* 3, 239–257.

Smilde, K. W. (1975). Micronutrient requirements of chrysanthemums grown on peat substrates. *Acta Hort* 50, 101–113.

Sonneveld, C. and Voogt, S. J. (1975). The effect of steam sterilisation of soil on the manganese uptake by glasshouse crops. *Acta Hort* 51, 311–319.

Taylor, J. B. and Joiner, J. N. (1963). Effects of three levels each of manganese, zinc and disodium ethylenebisdithiocarbamate on the growth and chemical composition of *Chrysanthemum morifolium* 'Iceberg'. *Proc Am Soc hort Sci* 83, 761–768.

Tsujita, M. J., Kiplinger, D. C., Tayama, H. K. and Staby, G. (1974). The effects of nitrogen source, temperature and light intensity on standard chrysanthemums. *Ohio Florists' Assn Bull* 541, 3–4.

Tsujita, M. J., Kiplinger, D. C., Tayama, H. K. and Staby, G. (1975). The effects of ammonium versus nitrate nitrogen on the growth, flowering, quality and chemical composition of standard chrysanthemum growth in midwinter. *Ohio Florists' Assn Bull* 551, 6–7.

Wadsworth, G. A. and Butters, R. E. (1972). The nutrition of A.Y.R. spray chrysanthemums in loamless medial *In* The Nutrition of Protected Crops. 2nd Colloq Potass Inst Ltd Kinsealy Dublin 1971, 83–88. *Henley-on-Thames Potass Inst Ltd.*

Waters, W. E. (1964). The effects of soil mixture and phosphorus on growth responses and phosphorus content of *Chrysanthemum morifolium. Proc Am Soc hort Sci* **84,** 588–594.

Waters, W. E. (1965). Influence of nutrition on flower production, keeping quality, disease susceptibility and chemical composition of *Chrysanthemum morifolium. Proc Am Soc hort Sci* **86,** 650–655.

Winsor, G. W. (1968). Studies of the nutrition of flower crops *Scient Hort* **20,** 26–40.

Winsor, G. W. (1972). The nutrition of glasshouse ornamental crops (non-woody). In The Nutrition of Protected Crops. 2nd Colloq Potass Inst Ltd Kinsealy Dublin 1971, 49–60. *Potass Inst Ltd Henley-on-Thames.*

Winsor, G. W. and Hart, B. (1965). Nutritional trials with glasshouse and other crops. *Rep Glasshouse Crops Res Inst 1964,* 65–70.

Winsor, G. W. and Hart, B. (1966). Nutritional trials with glasshouse crops. *Rep Glasshouse Crops Res Inst 1965,* 80–84.

Woltz, S. S. (1957). Studies on the nutritional requirements of chrysanthemums. *Proc Fla St hort Soc* **69,** 352–356.

Woltz, S. S. (1960). Symptoms of nutritional disorders of chrysanthemums and gladiolus. *Proc Fla St hort Soc* **72,** 383–385.

CHAPTER 8

Poinsettia

The Poinsettia *(Euphorbia pulcherrima)*, like the chrysanthemum, flowers in short days in response to a period of continuous darkness of 12 hours or more. The time of flowering can be controlled therefore by means of lighting and shading. Poinsettias need good light and a day temperature of at least 21°C to achieve their maximum rate of growth; night temperatures of 21°C or more delay flowering and day temperatures below 15.5°C inhibit growth. Many varieties need a minimum day temperature of 20°C to achieve full development of the bracts, but this temperature is then reduced to harden the plants for sale and to intensify the colouration of the bracts (Shanks, 1969). The current recommendations for controlled flowering in the UK are similar (Ministry of Agriculture, 1977), except that the temperature is reduced for three weeks (18°C in the day and 15°C at night) to stimulate flower initiation.

Poinsettias are susceptible to root and collar rot, hence the use of a well drained substrate is essential. Peat-sand composts have been recommended for this purpose, but fine peat should be avoided (Ministry of Agriculture, 1977). In order to maintain the nutrient status of such composts, it was suggested that a feed containing 200 mg l^{-1} N, 165 mg l^{-1} K and 90 mg l^{-1} P should be applied at every watering, commencing 2–3 weeks after potting and continuing until 2 weeks before marketing.

A large increase (over 50%) in the fresh weight of the plants is associated with the development of the bracts, accompanied by a decrease in the N and K contents of the growing medium (Boodley, 1969). Weekly applications of nutrient solution containing 400 mg l^{-1} NO_3-N and 400 mg l^{-1} K were necessary to maintain adequate nutrient levels for the crop.

Poinsettias grow best when the pH of the soil is in the range 4.5–6.5, and excessive liming results in deficiencies of P and Zn (Shanks, 1962). Acidity can, however, induce Mo deficiency (Royle, 1984). Tsutsui and Aoki (1981) found the best growth in sand culture at pH 5.0.

Nitrogen

Symptoms and growth responses. Plants grown without added N showed a uniform yellowing of all leaves, beginning at the bottom of the plant (Laurie and Wagner, 1940). Deficient plants first became uniformly pale green and then yellow, followed by loss of leaves from the base upwards (Widmer, 1953). Growth was retarded and

'hard', and leaf size was markedly reduced. The internodes were short and the red colouration of the petioles became more intense. Premature senescence of the leaves and delayed flowering were features of the deficiency. Shanks (1962, 1969) found that a deficiency also resulted in poor rooting of cuttings and reduced uptake of Mg due to poor root action. Tsutsui and Aoki (1981) found that a deficiency during the early stages of development retarded growth severely, and that subsequent applications of N did not restore normal growth. A deficiency in more mature plants resulted in yellowing of the lower leaves and a reduction in the size of the bracts.

Sprays containing urea improved the colour of the leaves of deficient plants, though insufficient N was supplied to restore the plants to normal growth (Meyer and Boodley, 1964). Weekly applications of sprays containing more than 6 g l^{-1} urea scorched the margins of the leaves.

Nitrogen deficiency reduced plant height, the dry weight and the number and size of the leaves and bracts (Tsutsui and Aoki, 1981); see also earlier studies by Shanks and Link (1957). High levels of NO_3-N tended to reduce stem length and dry weight, though not significantly (Gaffney *et al.*, 1982). Ammonium-nitrogen used as the only source of N reduced stem length by more than 50% and dry weight by over 60%. Increasing the ratio of NH_4-N to NO_3-N reduced stem length, the number of nodes and the dry weight of the plants. None of the plants receiving more than 50% of the total N as NH_4-N were marketable (Gaffney *et al.*, 1982). Previously, Boodley (1971) had shown that calcium nitrate gave much better growth than ammonium nitrate.

High levels of NH_4-N resulted in interveinal chlorosis of the lower leaves, the margins of which became scorched and rolled upwards and inwards (Boodley, 1971). These symptoms were sometimes preceded by wilting of the lower leaves and followed by shedding of the affected leaves. Gaffney *et al.* (1982) noted that plants receiving NH_4-N were stunted, the leaves and stems were brittle, and that there was little new growth. The lower leaves were yellow, the petioles were a darker red than normal and mild symptoms of K deficiency were induced on the younger leaves. The roots were short, unbranched and brown. Cox and Seeley (1984) reported interveinal chlorosis and marginal necrosis of the leaves of plants grown in solution culture at 50 mg l^{-1} NH_4-N in the presence of calcium carbonate.

Nitrogen content of the leaves. A critical level of 3.0% N in the leaves has been suggested (Matkin, quoted by Kofranek and Paul, 1978; Fortney and Wolf, 1981) and deficient plants were found to contain 2.0–2.4% N (Meyer and Boodley, 1964), 2.5% N (Holcomb and Williams, 1980) and 2.3% N (Tsutsui and Aoki, 1981). Normal growth is associated with 4.1% N (Meyer and Boodley, 1964), 3.0–4.7% N (Boodley, 1969), 4.5% N (Criley and Carlson, 1970), 4.7% (Boodley, 1980), 4.0–6.0 (Fortney and Wolf, 1981) and 4.8% N (Tsutsui and Aoki, 1981) in the leaves. Nine cultivars studied by Boodley (1977b) contained 4.7–5.6% N, with an average of 5.1% N. The content of the leaves increased from 2.6% N at low N to 3.3% at medium N and 4.5% at high N, accompanied by progressive increases in length of the bracts and the internodes and also in the number of flowering shoots per plant (Shanks and Link, 1957). Fortney and Wolf (1981) regarded 7.3% N in the leaves as excessive.

Phosphorus

Deficiency symptoms and growth responses. Poinsettias grown without added P made little growth. The older leaves yellowed progressively inwards from the margins and abscissed (Laurie and Wagner, 1940). Phosphorus deficiency restricted growth, and the upper leaves became very dark green in colour. The lower leaves became yellow and died prematurely and progressively up the plant (Widmer, 1953; see also Koths et al., 1980). The tissue between the veins may become puckered and rough in texture (Struckmeyer, 1960) and brown areas may form on the yellow leaves (Tsutsui and Aoki, 1981). Poor rooting of cuttings and delayed flowering have also been noted (Shanks, 1962, 1969).

Plant height, dry weight, the number of leaves formed and the total leaf area were greatly reduced by a deficiency of P (Tsutsui and Aoki, 1981).

Phosphorus content of the leaves. Matkin, quoted by Kofranek and Paul (1978), reported a critical level of 0.2% P in the leaves. Shanks and Link (1957) found contents of 0.27% P and 0.19% P in the leaves of plants grown without added P in a soil:peat substrate; the corresponding values with added P were 0.37% and 0.34%. Meyer and Boodley (1964) reported 0.10% and 0.19% P in the leaves of plants grown without added P.

Normal plants were found to contain 0.54% P (Meyer and Boodley, 1964) and 0.28–0.48% P (Seeley, 1977), and ranges of 0.35–0.97% P (quoted by Criley and Carlson, 1970) and 0.30–0.70% P (Fortney and Wolf, 1981) have been suggested as typical. Concentrations above 0.7% P have been regarded as excessive (Fortney and Wolf, 1981), and although Boodley (1969) found that the leaves of healthy plants contained 0.80–1.40% P during the first twelve weeks after planting, the latter value (1.4% P)

seems abnormally high. Boodley (1977b) reported values of 0.82–1.06% P for a group of nine cultivars, and 0.47–0.51% P in an experiment comparing different sources of trace elements (Boodley, 1980). Tsutsui and Aoki (1981) recorded contents of 0.14–1.79% P in the leaves of poinsettias grown in sand culture over the range 0–4 mg l^{-1} P. The highest of these concentrations proved excessive, with significant reductions in plant height and in bract diameter.

Potassium

Symptoms and growth responses. Laurie and Wagner (1940) observed yellowing of the margins of the lower leaves, spreading interveinally and at first leaving the veins dark green. The leaf margins later turned brown, but the leaves remained attached to the plant for some time after dying. All leaves were eventually affected except the very young leaflets at the top of the plants. Stuart and Rocke (1951) described yellowing of the older leaves, and noted that the terminal portions of the leaves were scorched. The leaves usually rolled inwards, and small, irregular tan-coloured spots developed between the veins at the tips and outer edges. Widmer (1953) recorded yellowing of the leaf margins as normally the first visible symptom of K deficiency, followed by the development of small brown spots ('flecking'); in some instances, however, the brown spots developed along the margins of dark green leaves in the middle of the plant. Eventually, the margins of leaves with symptoms curled downward and these leaves dropped off. Struckmeyer (1960) reported stunted growth, shortening of the internodes, interveinal mottling and, eventually, drying out of the margins of the affected leaves. Shanks (1962, 1969) noted 'pin point speckling' of the leaves, usually at the time of flowering. In more severe instances of deficiency, however, scorching of the leaves resulted in defoliation just before flowering. Meyer and Boodley (1964) reported bronzing and down-curling of the leaves of deficient plants. Koths et al. (1980) described interveinal chlorosis of the older leaves, progressing to necrosis and advancing up the plant. The illustrations of Tsutsui and Aoki (1981) show the development of yellow or pale brown spots near the margins of the mature leaves. The spots spread inwards between the veins and the surrounding tissue became yellow.

Link and Shanks (1957) found that the number of rooted cuttings produced per stock plant, and the average degree of rooting, increased with the level of K application. Boodley (1977a) found no significant effects of concentration over the range 0–500 mg l^{-1} K on plant height, bract size, fresh or dry weight for plants grown in a peat:vermiculite substrate (Cornell peat-lite mix A), but warned of the need to supply K when using peat: perlite (peat-lite mix B). Similar results were obtained by Seeley (1977); it was concluded that the vermiculite was supplying some K to the crops. In sand and solution culture, deficiency greatly reduced plant height, dry weight and the number of bracts but had no effect on the number of leaves formed (Tsutsui and Aoki, 1981). A high level of K depressed plant height and the number of leaves formed, but had no effect on dry weight or on the number of bracts per plant.

Potassium content of the leaves.   The leaves of deficient plants contained less than 1.0% K (Matkin, quoted by Kofranek and Paul, 1978), 0.45–0.78% K (Meyer and Boodley, 1964) and 0.5% K with a lower critical level of 1.3% K (Tsutsui and Aoki, 1981). Shanks and Link (1957) recorded concentrations of 2.89, 3.45 and 3.77% K in the leaves of poinsettias grown in a loam:peat substrate with five applications of solutions containing 138, 415 and 1246 mg l^{-1} K. Boodley (1969) found 2.0–3.8% K in the leaves of healthy plants during the first twelve weeks after planting whilst Criley and Carlson (1970) quoted a range of 2.1–2.8% K for good growth. Seeley (1977) suggested that 1.7–2.5% K was satisfactory for recently matured leaves, or up to 3.1% K in lower leaves. Tsutsui and Aoki (1981) found 2.4–3.7% K in the healthy leaves. A range of 1.5–3.5% K was also given for normal plants by Fortney and Wolf (1981), 4.0% K being regarded as excessive, though Meyer and Boodley (1964) found 3.9–4.6% K in healthy plants.

Calcium

Deficiency symptoms and growth responses.   Laurie and Wagner (1940) noted that the uppermost leaves of Ca deficient plants became abnormally dark green with a reddish tint, and tended to wilt. The stems were stiff, the terminal buds died and complete defoliation followed. Stuart and Rocke (1951) observed small, puckered leaves which curled downwards. Omission of Ca was the only nutritional treatment to cause early death of the plants. Calcium deficiency resulted in spindly growth, short internodes, and dark green foliage with intensely dark red petioles (Widmer, 1953). When Ca was omitted entirely, the upper leaves became limp and curled upwards whilst the lower leaves became yellow and died prematurely. Development of brown areas on the upper leaves, which became 'cupped', and arrested development of the growing point were noted by Struckmeyer (1960). Tsutsui and Aoki (1981) illustrated chlorosis of the leaf margins, gradually spreading inwards to become almost cream in colour. The edges of the leaves curled downwards, and pale brown patches developed.

Koths *et al.* (1980) recorded slow growth, sometimes resulting in poor bract formation. Calcium deficiency reduced the dry weight of the plants, the number and size of the bracts and tended to reduce plant height, but had no effect on the number and size of the leaves (Tsutsui and Aoki, 1981).

Calcium content of the leaves.   A critical level of 0.5% Ca in the leaves has been proposed by Matkin, quoted by Kofranek and Paul (1978), and by Fortney and Wolf (1981). Tsutsui and Aoki (1981) found 0.5% Ca in the leaves of deficient poinsettias.

Shanks and Link (1957) recorded values of 0.63–0.72% Ca in the leaves of healthy plants, increasing progressively with the level of N applied. Link and Shanks (1957) found 0.9–1.5% Ca in the leaves of poinsettia stock plants. Meyer and Boodley (1964) reported 0.66–0.83% Ca in the leaves of plants receiving a complete nutrient solution, falling to 0.23–0.35% Ca where N was omitted. Boodley (1969) found 1.1–2.6% Ca with an average of 1.6% Ca. Criley and Carlson (1970) quoted optimal values of 0.74% and 1.5% Ca, the former from a commercial source. Seager (1973) reported far higher values of 3.3–3.7% Ca in the leaves. Values tabulated by Holcomb and Williams (1980) from trials with various slow-release (NPK) fertilisers showed a range of 0.71–1.38% Ca in the leaves. Boodley (1977b) reported values of 1.07–1.35% Ca for nine cultivars studied, with a mean value of 1.23% Ca. Fortney and Wolf (1981) regarded values of 0.7–2.0% Ca as normal. Tsutsui and Aoki (1981) found contents of 0.5–2.9% Ca in the leaves of poinsettias grown in sand culture at five concentrations of Ca (0–80 mg l^{-1} Ca). Plants containing 1.5–1.8% Ca (interpolated values) appeared to give the best growth, with a 32% reduction in total bract area at the highest level (2.9% Ca) in leaves. The calcium content of the plants was depressed by high levels of nitrogen, potassium and magnesium and by low levels of phosphorus.

Magnesium

Deficiency symptoms and growth responses.   Laurie and Wagner (1940) observed severe stunting of Mg deficient poinsettias, with interveinal chlorosis of the lower leaves accompanied by puckering of the laminae. Necrotic spots developed along the margins and between the veins, and the leaves eventually died. In addition to chlorosis of the older leaves, Stuart and Rocke (1951) particularly noted the large dead spots which developed a short distance from the edges of the leaves. These spots formed a regular pattern, quite unlike the smaller and irregular spots associated with K deficiency. Growth was restricted by Mg deficiency; the foliage became pale green and interveinal yellowing developed on the lower leaves (Widmer, 1953). Later, sandy coloured spots or small sandy areas developed interveinally; these rarely started at the margin and were not always preceded by yellowing, but led to a rapid loss of the affected leaves. Struckmeyer (1960) noted puckering and yellowing of the interveinal tissue. The main veins remained green and very prominent, and the leaf margins sometimes became 'cupped'. Dickey (1977) published a photograph showing chlorosis and necrosis of field-grown poinsettias. The illustrations of Tsutsui and

Aoki (1981) show a somewhat uneven development of interveinal yellowing on the lower leaves; the colour deepened to orange-yellow as the symptoms became more severe. Magnesium deficiency had no effect on plant height or on the total number of leaves and bracts formed, but slightly reduced the dry weight of the plants and increased the number of leaves that died prematurely (Tsutsui and Aoki, 1981). Addition of Mg to a soil-peat substrate increased the number of rooted cuttings obtained per poinsettia stock plant, and also the average degree of rooting (Link and Shanks, 1957).

Magnesium content of the leaves. Link and Shanks (1957) recorded 0.27–0.39% Mg in the leaves of poinsettia stock plants grown without added Mg, and 0.54–0.80% Mg, accompanied by improved yields of rooted cuttings, where this nutrient was supplied. Plants grown in a complete nutrient solution contained 0.42–0.43% Mg in their leaves (Meyer and Boodley, 1964). Boodley (1969) recorded values of 0.6–1.3% Mg in the leaves of healthy plants, with an average of 0.81% Mg. Criley and Carlson (1970) quoted optimal values of 0.57% and 0.6% Mg from two different sources, whilst Seager (1973) gave a range of 1.1–1.3% Mg. Boodley (1977b), comparing nine poinsettia cultivars, found values of 0.50–0.72% Mg in the leaves, with an average of 0.62% Mg. Kofranek and Paul (1978) quoted lower critical levels of 0.11%, 0.15% and 0.2% Mg from three sources. Values of 0.60%–0.82% Mg were recorded by Holcomb and Williams (1980) from their trials with slow-release fertilisers. Fortney and Wolf (1981) regarded values below 0.2% Mg as indicating deficiency, with a normal range of 0.4–1.0% Mg. Tsutsui and Aoki (1981) found 0.20% Mg in the leaves of deficient plants, and gave a critical level of 0.45% Mg: a provisional figure of 0.9% Mg was indicated as possibly excessive. The leaf Mg content increased with the levels of Mg and P applied but was depressed by increasing levels of K and Ca.

Iron

Little has been published concerning Fe deficiency in poinsettias. Koths *et al.* (1980) described the young leaves as chlorotic with green veins, symptomatic of root damage. In their general description of this deficiency in ornamental plants, however, it was noted that the veins of the young leaves slowly became white, unlike Mn deficiency where they remain green.

A critical level of 50 μg g^{-1} Fe in the leaves was quoted by Kofranek and Paul (1978). A range of 100–500 μg g^{-1} Fe in the leaves was suggested for normal plants (Fortney and Wolf, 1981), with deficiency below 50 μg g^{-1} Fe. Iron contents in the range 146–386 μg g^{-1} (Boodley, 1969), 150–156 μg g^{-1} (Criley and Carlson, 1970) and 58–166 μg g^{-1} (Holcomb and Williams, 1980) were found in healthy plants; Boodley (1977b) found 167–381 μg g^{-1} Fe in a series of nine cultivars. The diversity of these values suggests that the total content is not a good indicator of the physiologically active iron content of the leaves: for further discussion of this point, see under Tomato (p. 62).

Manganese

The symptoms of Mn deficiency were described by Laurie and Wagner (1940). A chlorosis developed at the tops of the plants, the leaves having a more or less netted appearance. The affected plants were about half the size of those receiving Mn, and had smaller bracts.

Critical leaf values of 30 μg g^{-1} and 45 μg g^{-1} Mn were reported by Kofranek and Paul (1978). Fortney and Wolf (1981) regarded values below 30 μg g^{-1} Mn as indicating deficiency. The leaves of healthy plants were found to contain 126–240 μg g^{-1} Mn (Boodley, 1969), 75–150 μg g^{-1} (Criley and Carlson, 1970), 46–89 μg g^{-1} (Holcomb and Williams, 1980) and 120–287 μg g^{-1} (Boodley, 1977b). A range of 100–200 μg g^{-1} Mn was suggested for normal plants, with 250 μg g^{-1} or more corresponding to toxicity (Fortney and Wolf, 1981). Higher concentrations of Mn (413–727 μg g^{-1}) were associated with a depression in the height and weight of the plants (Boodley, 1980).

With excessive Mn levels brown spots appeared on the older leaves and spread to the younger ones, and there was a decrease in growth. Symptoms of toxicity first became apparent when the leaves contained 3440 μg g^{-1} Mn (Coorts, 1958). This is a relatively high value, to be seen in conjunction with data for other crops in this study including chrysanthemums (2670 μg g^{-1} Mn), described as 'non-tolerant', and carnations (5700 μg g^{-1} Mn), referred to as tolerant.

Copper

Copper deficiency is associated with concentrations of less than 5 μg g^{-1} Cu in the leaves (Fortney and Wolf, 1981). Matkin, quoted by Kofranek and Paul (1978), suggested a relatively low value of 1 μg g^{-1} as the critical level. Fortney and Wolf (1981) associated normal growth with 6–15 μg g^{-1} Cu, and the values reported generally fall within this range; 8 μg g^{-1} (Criley and Carlson, 1970), 8.4–13.4 μg g^{-1} (Boodley, 1977b) and 8.0–12.3 μg g^{-1} (Boodley, 1980). Of the 14 values reported by Holcomb and Williams (1980) in the range 3.1–9.1 μg g^{-1} Cu, nine were below 4.5 μg g^{-1} Cu yet no reference was made to deficiency symptoms. Boodley (1969) had previously found values down to 4.0–4.5 μg g^{-1} Cu within his overall range of 4.0–12.5 μg g^{-1} Cu without reference to any symptoms of deficiency. It thus appears that identifiable symptoms of this deficiency in poinsettias may not result until the leaves contain less than 3–4 μg g^{-1} Cu.

Zinc

Struckmeyer (1960) recorded that symptoms of Zn deficiency first appeared on the upper leaves, beginning with an interveinal mottling. The leaves became long and narrow, and the midribs grew more rapidly than the adjacent tissue, thus causing buckling. Dry, necrotic areas later developed in the leaves. Growth of the plants was only slightly restricted. Interveinal chlorosis due to this deficiency was illustrated by Shanks (1962). The condition was induced by excessive liming, and could readily be controlled by drenching the soil with the Zn – containing fungicide, Zineb.

A critical level of 15 μg g^{-1} Zn in the leaves was proposed by Matkin, quoted by Kofranek and Paul (1978), and a similar figure was associated with deficiency by Fortney and Wolf (1981). Healthy plants contained 36–94 μg g^{-1} Zn (Boodley, 1969), 25–60 μg g^{-1} (Fortney and Wolf 1981), 31–150 μg g^{-1} (Criley and Carlson, 1970, from two sources) and 29–57 μg g^{-1} (Holcomb and Williams, 1980); Boodley (1977b) found 72–172 μg g^{-1} Zn in a range of cultivars.

Boron

Deficiency symptoms. Laurie and Wagner (1940) induced B deficiency in sand culture. After three months the terminal buds ceased to grow, and the lateral buds near the tips developed. The terminal leaves thickened and tended to roll from the tip to the base. Some of the midribs cracked on the undersides of the leaves. The bracts developed slowly and abnormally. Stuart and Rocke (1951) recorded that the upper leaves became somewhat stiff and that the petioles cracked open, allowing sap to exude. Overall growth was not, however, greatly reduced. Widmer (1953) described yellowing of the leaves of deficient plants, beginning at the margins and spreading interveinally. The growing points rapidly became distorted, and new foliage was reduced in size. The edges of the leaves curled downwards and the petioles became dark purple in colour. The foliage was very brittle and droplets of latex were exuded from the undersides of the leaves particularly along the large veins and near the base of the leaves. Struckmeyer (1960) reported that the very short internodes gave the plants a rosette-like appearance; failure of the terminal growing point was accompanied by the development of axillary shoots from the upper nodes.

Boron content of the leaves. A critical level of 20 μg g^{-1} B in the leaves was quoted by Kofranek and Paul (1978); see also Fortney and Wolf (1981). The latter authors also indicated a range of 30–100 μg g^{-1} B for healthy plants, and many of the published values lie within this range. Thus Boodley (1969) reported 32–49 μg g^{-1} B for twelve leaf samples taken at weekly intervals. Criley and Carlson (1970) quoted optimal levels of 28 and 100 μg g^{-1} B. Boodley (1977b) found 30–57 μg g^{-1} B in a group of nine

cultivars, and Holcomb and Williams reported 19–62 μg g^{-1} B from a study involving 14 fertiliser treatments. Data from further trials by Boodley (1980) showed a range of 26–34 μg g^{-1} B in the leaves.

Effects of excess boron. Concentrations of 200 μg g^{-1} B or more were regarded as excessive (Fortney and Wolf, 1981). No symptoms of toxicity were reported for recently matured leaves containing 150–180 μg g^{-1} B (Kofranek et al., 1956) but the tips of leaves containing 400–626 μg g^{-1} B (upper and lower leaves respectively) were scorched. With higher B contents, interveinal yellowing developed on the lower leaves, followed by marginal scorching before the leaves were shed. The degree of leaf abscission by plants grown in sand culture at four concentrations was 5% at 0.3 and 2.3 mg l^{-1} B, 25% at 4.8 mg l^{-1} B and 45% at 10.3 mg l^{-1} B. Levels of up to 1664 μg g^{-1} B were recorded in the lower leaves of plants receiving excess B. Toxicity reduced the size of the bracts by 11% but had no effect on the time of flowering.

Molybdenum

The symptoms of Mo deficiency, namely yellowing and browning of the leaves, have been described by Jungk et al. (1970). These were most acute when a high level of N was supplied. Damage was reduced by supplying Mo in the compost, by liming to pH 6–7 and by avoiding high levels of N. A spray containing 100 mg l^{-1} Mo restored plants with a mild deficiency to healthy growth. Royle (1984) noted that poinsettias were particularly susceptible to this deficiency which first shows as a yellowing of the edges of the younger leaves, spreading to form pale areas between the veins. These pale areas later shrivel, making the plants unsaleable. The condition can be avoided by liming to eliminate acidity in the substrate, incorporating a suitable fritted trace element mixture and spraying the foliage with either sodium or ammonium molybdate (0.25 g l^{-1}) three weeks after potting.

Few analytical data are available for poinsettia, but 1.3–1.7 μg g^{-1} Mo were found in healthy leaves (Boodley, 1980).

References

Boodley, J. W. (1969). Nutrient content of 'Paul Mikkelsen' poinsettias from juvenile to mature growth. *N Y St Flow Gro Bull* **288**, 1–4, 7.

Boodley, J. W. (1971). Nitrogen fertilisers and their influence on the growth of poinsettias. *N Y St Flow Ind Bull* **10**, 4–7.

Boodley, J. W. (1977a). Poinsettia nutrition studies, 1976: levels of potassium fertilisation. *N Y St Flow Ind Bull* **88**, pp. 3, 4, 7.

Boodley, J. W. (1977b). Some effects of three trace element fertilisers on the growth of nine cultivars of poinsettias. *N Y St Flow Ind Bull* **89**, pp. 3, 4, 7, 8.

Boodley, J. W. (1980). Comparison of four trace element fertiliser sources in peat-lite mixes. *Acta Hort* **99**, 33–38.

Coorts, G. D. (1958). Excess manganese nutrition of ornamental plants. *Bull Mo agric Exp Stn* **669**, 35 pp.

Cox, D. A. and Seeley, J. G. (1984). Ammonium injury to Poinsettia: effects of NH_4-N:NO_3-N ratio and pH control in solution culture on growth, N absorption and N utilisation. *J Am Soc hort Sci* **109**, 57–62.

Criley, R. A. and Carlson, W. H. (1970). Tissue analysis standards for various floricultural crops. Criley, R. A. and Carlson, W. H. (1970). Tissue analysis standards for various floricultural crops. *Florists' Rev* **146**, 19, 20, 70–73.

Dickey, R. D. (1977). Nutritional deficiencies of woody ornamental plants used in Florida landscapes. *Bull Fla agric Exp Stn* **791**, 63 pp.

Fortney, W. R. and Wolf, T. K. (1981). Determining nutritional status: Plant Analysis. *Penn Flow Gro Bull* **331**, 1, 5–11.

Gaffney, J. M., Lindstrom, R. S., McDaniel, A. R. and Lewis, A. J. (1982). Effect of ammonium and nitrate nitrogen on growth of poinsettia. *HortSci* **17**, 603–604.

Holcomb, E. J. and Williams, S. (1980). The effect of slow-release fertilisers on poinsettias. *Penn Flow Gro Bull* **324**, 1, 6, 7.

Jungk, A., Malaheb, B. and Wehrmann, J. (1970). Molybdenum deficiency of poinsettia, a cause of leaf damage. *Gartenwelt* **70**, 31–33, 35.

Kofranek, A. M., Lunt, O. R. and Kohl, H. C. (1956). Tolerance of poinsettias to saline conditions and high boron concentrations. *Proc Am Soc hort Sci* **68**, 551–555.

Kofranek, A. M. and Paul, J. L. (1978). Critical nutrient levels for some selected floricultural crops. *Calif agric Exp Stn Bull* **1879**, 32.

Koths, J. S., Judd, R. W., Maisano, J. J., Griffin, G. F., Bartok, J. W. and Ashley, R. A. (1980). Nutrition of greenhouse crops. *Bull Connecticut Greenhouse Crop Production Task Force and Coop Ext Ser Northeast States, U.S.A.* 16 pp.

Laurie, A. and Wagner, A. (1940). Deficiency symptoms of greenhouse flowering crops. *Bull Ohio agric Exp Stn* **611**, 26 pp.

Link, C. B. and Shanks, J. B. (1957). The mineral nutrition of Poinsettia stock plants in the greenhouse. *Proc Am Soc hort Sci* **69**, 502–512.

Meyer, M. M., Jr. and Boodley, J. W. (1964). Foliar applications of nitrogen, phosphorus and potassium to chrysanthemum and poinsettia. *Proc Am Soc hort Sci* **84**, 582–587.

Ministry of Agriculture, Fisheries and Food (1977). Poinsettia *Euphorbia pulcherrima* (Wild). Adv leaf 610. *HMSO Lond.*, 12 pp.

Royle, S. (1984). Treat your poinsettias with molybdenum. *In South Coast Glasshouse Advisory Unit, Chichester,* Technical Note **91**, 8–9.

Seager, J. C. R. (1973). Effect of peat type and frequency of watering on Poinsettia. *Meded Fac Landbouwwet Rijksuniv Gent* **38**, 2038–2045.

Seeley, J. G. (1977). Potassium fertilisation of poinsettias in two peat-lite mixes. *Florists' Rev* **161**, 31–33, 75, 76.

Shanks, J. B. (1962). Poinsettias, their culture. *Maryl Florist* **89**, 32 pp.

Shanks, J. B. (1969). Poinsettias – greenhouse culture. *Maryl Florist* **152**, 27 pp.

Shanks, J. B. and Link, C. B. (1957). The mineral nutrition of poinsettias for greenhouse forcing. *Proc Am Soc hort Sci* **69**, 502–512.

Struckmeyer, B. E. (1960). The effect of inadequate supplies of some nutrient elements on foliar symptoms and leaf anatomy of poinsettia *(Euphorbia pulcherrima* Willd.). *Proc Am Soc hort Sci* **75**, 739–747.

Stuart, N. W. and Rocke, R. M. (1951). Nutrient deficiency effects on poinsettias. *Florists' Exchange* **1117** 10–11.

Tsutsui, K. and Aoki, M. (1981). Response of poinsettias to major nutrient supply in relation to nutrient uptake and growth. *Bull Veg Ornam Crops Res Stn Anô Jap Ser A* **8**, 171–207.

Widmer, R. E. (1953). Nutrient studies with the poinsettia. *Proc Am Soc hort Sci* **61**, 508–514.

Colour Illustrations

Cucumber

Nitrogen deficiency

Growth is restricted and fewer fruits develop. The new leaves are small and pale yellow-green in colour whilst the older leaves become yellow and die prematurely. The fruit are short and pale green.

1 The leaves of deficient plants (bottom of picture) are reduced in size and become yellow-green or yellow in colour (cv. Femdam). Leaf analysis (laminae only for this crop except when stated otherwise), 5.0% and 2.0% N in the normal and deficient leaves respectively; deficiency, <2.5% N; normal range, 3.5 – 5.5% N.

2 Restriction of the growth becomes more obvious as the severity of the deficiency increases (cv. Brilliant).

Phosphorus deficiency

Growth and fruit production are seriously restricted. The young leaves are small and dull green in colour. Brown interveinal areas develop on the mature leaves, and the oldest leaves die prematurely.

3 New growth is restricted and deficiency symptoms develop on the mature leaves (front row of plants; cv. Uniflora D).

4 The yellow interveinal areas, which later become brown, are randomly distributed over the leaf surface (cv. Brilliant). Leaf analysis, 0.2% P; deficiency, <0.25% P; normal range, 0.35 – 0.8% P.

5

7

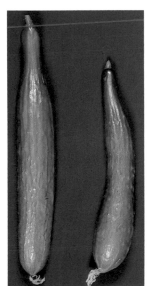

9

6

8

5 The brown areas on the lower leaves enlarge rapidly, the margins may also become scorched and the leaves die prematurely (cv. Brilliant). Leaf analysis, 0.1% P.

Potassium deficiency

Plant growth may not be affected seriously but the yield of fruit may be greatly reduced.

6 A yellow margin develops on the expanded leaves and spreads interveinally; these leaves tend to curl downwards (cv. Farbio).

7 The margins of the middle and of the upper expanded leaves become yellow and may scorch rapidly in hot weather (cv. Uniflora D). Leaf analysis, 0.8% K in the laminae (3.3% K in the petioles); deficiency, <2.0% K (<10% K in the petioles); normal range, 3–5% K (12–16% K in the petioles).

8 When the deficiency becomes severe, the expanding leaves are also affected. The yellow margin becomes very broad and spreads interveinally. Purple areas, which rapidly become brown, may develop near the leaf margins (cv. Farbio). Leaf analysis, 0.7% K (3.5% K in the petioles).

9 Fruit from K deficient plants tends to be poorly developed at the stem end and somewhat swollen at the tip; a normal fruit has been included for comparison.

10

Calcium deficiency

The margins of the very youngest leaves become scorched and the expanding leaves become 'cupped'; in acute cases, the growing point may die. Milder symptoms of this type in the leaves may arise from restricted translocation of Ca due to high humidity or a dry substrate.

10 The margins of the youngest leaves are scorched and they develop a 'cupped' shape as they expand (cv. Brilliant).
Healthy leaves usually contain 2–10% Ca.

12

11

13

Magnesium deficiency

Interveinal yellowing develops on the middle and lower leaves and gradually progresses up the plant.

11 Yellow interveinal areas develop on the mature leaves and spread up the plant. The yellowing spreads across the leaf surface until only the main veins remain green (cv. Brilliant).
Leaf analysis, 0.16% Mg; deficiency, <0.30% Mg; normal range, 0.4–0.8% Mg.

12 Leaves on the main stem may develop very pale brown or white interveinal areas which appear to radiate in bands from the petiole (cv. Farbio).
Leaf analysis, 0.20% Mg.

13 Leaves on the laterals develop yellow interveinal areas which may rapidly become pale brown; the margins as well as the main veins remain green at first (cv. Brilliant).

14

Iron deficiency

The youngest leaves are yellow and reduced in size; they may eventually become almost white if the deficiency is not corrected. A temporary deficiency may be induced by overwatering (poor aeration) or by the loss of roots due to pests or disease.

14 The young leaves become pale yellow as the deficiency progresses. The older leaves become pale green, and develop large pale brown areas before they collapse and die prematurely (cv. Telegraph).
Leaf analysis, 42 μg g^{-1} Fe. Estimates of the total Fe content of the foliage are frequently misleading, but values below about 80 μg g^{-1} Fe suggest the possibility of deficiency.

15

Manganese deficiency

The leaves become yellow-green in colour and growth is restricted.

15 The margin and interveinal parts of the leaf become pale green, later yellow-green or yellow. The fine veinal network remains green, however, and this characteristic is useful in distinguishing between Fe and Mn deficiencies (cv. Femspot). Leaf analysis, 7 μg g^{-1} Mn; deficiency, <20 μg g^{-1} Mn; normal range, 100–300 μg g^{-1} Mn.

16

Manganese toxicity

Excess manganese gives rise to reddish-black deposits in the veins of the lower leaves, and seriously affected leaves die prematurely. See also the general note on this disorder under Tomato (p. 140).

16 Yellowing develops around the veins, spreads interveinally and the affected leaves die prematurely (cv. Femdam). Leaf analysis, 3200 μg g^{-1} Mn; toxicity, >900 μg g^{-1} Mn; normal range, 100–300 μg g^{-1} Mn.

17

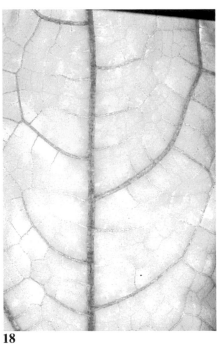

18

Manganese toxicity continued

17 The main veins become reddish-black due to the accumulation of Mn compounds (cv. Fcmdam).

18 Deposits of Mn compounds around the veins of the leaves are seen most easily when viewed by transmitted light.

19 Excess Mn also accumulates in the form of small, purple coloured spots at the base of the spines (trichomes) on the petioles and undersides of the leaves (compare with the normal leaf, plate 20). This symptom is specific to Mn toxicity (cv. Femdam).

20 The petiole and underside of a normal green leaf are covered with small white spines (cv. Femdam); compare with leaf affected by excess Mn (plate 19).

19

20

21

Copper deficiency

Growth is restricted, the leaf size is greatly reduced and few fruit develop. The appearance of the leaves differs according to their position on the plant. The mature leaves develop a mottled appearance whilst the oldest leaves become yellow (interveinally) and die prematurely.

21 Yellowing on the margins of the oldest leaves spreads interveinally, later becoming pale brown; the affected leaves die prematurely (cv. Uniflora D). Leaf analysis, $<2\,\mu g\,g^{-1}$ Cu; deficiency, $<4\,\mu g\,g^{-1}$ Cu; normal range, $7–17\,\mu g\,g^{-1}$ Cu.

22 Yellowing first develops around veins near the margin of the leaf and then spreads towards the middle of the lamina. The margin remains green at first (cv. Femdam).

23 Yellowing around all the minor veins of mature leaves results in a mottled appearance; the main veins, parts of the margin and isolated interveinal areas remain green (cv. Femdam).

24 Flowering is restricted and the few fruits produced are poorly developed with small, sunken brown areas scattered over the yellow-green skin (cv. Femdam).

22

23

24

25

Zinc deficiency

Growth is restricted and the leaves, which are reduced in size, become yellow-green or yellow.

25 The leaves become yellow-green or yellow except for the main veins, which remain dark green and well defined. Parts of the margin eventually become pale brown (cv. Sporu).
Leaf analysis, 18 $\mu g\,g^{-1}$ Zn; deficiency, <25$\mu g\,g^{-1}$ Zn; normal range, 40–100 $\mu g\,g^{-1}$ Zn.

26

Excess zinc

26 The interveinal areas become yellow-green. The entire veinal network is at first dark green, later becoming somewhat blackened (cv. Bedfordshire Prize).

27 The blackish discolouration around the main veins is particularly prominent, a feature that helps to distinguish this disorder from Mn deficiency in which the veins remain green (see Plate 15).
Leaf analysis, 950 $\mu g\,g^{-1}$ Zn.

When this disorder is very severe, the young leaves become yellow and the symptoms resemble those of Fe deficiency (see Roorda van Eysinga and Smilde, 1981).

Boron deficiency

The foliage becomes very brittle and a broad cream margin develops on the oldest leaves. The growing point may die if the deficiency becomes severe.

28 Yellowing of the lower leaves spreads from the tip (top of picture) and develops into a broad cream band (bottom of picture) around the entire margin (cv. Femdam).
Leaf analysis, 8 $\mu g\,g^{-1}$ and 6 $\mu g\,g^{-1}$ B (top and bottom respectively); deficiency, <20 $\mu g\,g^{-1}$ B; normal range, 30–80 $\mu g\,g^{-1}$ B.

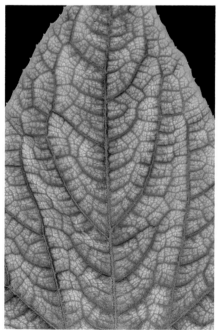

27

28

29

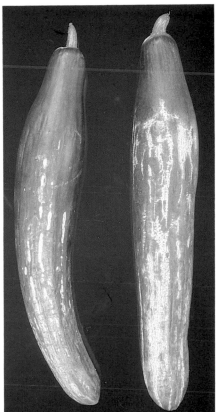

31

Boron deficiency continued

29 The broad cream margin, which develops on the lower leaves, progresses up the plant and the tips of the affected leaves eventually become brown, curling downwards and inwards (cv. Femdam).

30 New but deformed growth at the top of a B deficient cucumber plant following death of the primary growing point; note the small, crinkled new leaves in contrast to the larger, chlorotic foliage beneath.

31 Fruit may develop corky streaks which spread longitudinally; this symptom resembles the response to inadequate heating (cv. Femdam).

30

32

33

34

Boron toxicity

The margins of the leaves become scorched and expanding leaves become 'cupped'.

32 A narrow yellow margin, which rapidly becomes scorched (pale brown), develops on most of the leaves. Young, expanding leaves become 'cupped' once the margins are scorched (cv. Femdam).
Leaf analysis, 270 μg g^{-1} B; toxicity, >150 μg g^{-1} B; normal range, 30–80 μg g^{-1} B.

33 As the toxicity becomes severe the veins near the margin become very pale yellow. This symptom gradually spreads across the leaf (cv. Farbio).

34 Acute toxicity affects even the youngest leaves (cv. Farbio).

35

36

37

Molybdenum deficiency

Growth is restricted and the leaves become pale green. Small, pale yellow or white areas develop interveinally near the margins and spread inwards across the surface of the younger leaves. The lower leaves may wilt and die prematurely.

35 The foliage of young plants grown under acid conditions (pH <5) becomes pale green (right; cv. Femdam).

36 The margins, and later the interveinal areas, of the leaves become pale brown (cv. Femdam).

37 As the plants mature, the leaves remain pale yellow-green, with prominent veins and a broad, pale margin (cv. Femdam).

38

39

Molybdenum deficiency continued

38 When the deficiency develops on mature plants, small pale yellow areas develop interveinally near the margins of the upper leaves. The yellowing, which may become almost white, spreads firstly around the leaf margin and then interveinally towards the centre of the leaf (cv. Femdam). Leaf analysis, $0.1 \, \mu g \, g^{-1}$ Mo. Leaves from healthy plants contained 1.3–$2.2 \, \mu g \, g^{-1}$ Mo.

39 In some instances, parts of the leaves of mature plants remain green at first, giving them a blotchy appearance (cv. Femdam). See also plate 62 in Volume 1.

Other disorders

40 *Beet pseudo-yellows virus* The veins of the leaves remain dark green whilst sharply defined yellow areas develop between them. These symptoms could be mistaken for a nutritional disorder.

40

Lettuce

41

42

Nitrogen deficiency

Growth and development are restricted and the leaves become pale green or yellow-green in colour.

41 Growth is restricted by N deficiency (left) and the plants fail to form hearts (cv. Neptune). Plant analysis, 2.2% and 4.8% N in the deficient (left) and normal plants respectively; deficiency, <2.5% N; normal range, 3.5 – 5.5% N. The N content of winter lettuce varies with light intensity; it is generally higher in crops maturing in poor light (December and January) and lower in crops maturing under better light conditions (February and March).

42 The leaves are yellow-green in colour and the outer leaves may become yellow and die prematurely (cv. Ostinata).

43a

43b

Phosphorus deficiency

Growth and development are restricted, but the foliage usually remains green (cf. N deficiency).
43 At low levels of P (43a), growth and development are seriously restricted. Differences in growth reflect local variations in the available P content of the soil. High levels of P result in very uniform growth (43b; cv. Neptune).

44 In summer crops, the outer leaves may become yellow-green. Large brown areas develop interveinally and spread towards the leaf margins (cv. Ostinata). Plant analysis, 0.08% P in the outer leaves; whole plants contained 0.15% P compared with 0.58% P in healthy plants. Deficiency, <0.2% P; normal range 0.5–0.8% P.

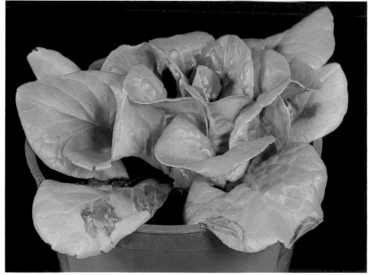

44

45

46

Potassium deficiency

Growth may not be affected seriously but the outer leaves become yellow as the plants mature.

45 Foliar symptoms of K deficiency may not appear until the plants are mature. The outer leaves then become yellow and may collapse rapidly in warm weather; these leaves easily become infected with *Botrytis cinerea;* (cv. Neptune).
Plant analysis, 2.2% K; deficiency, <2.5% K; normal range, 5–10% K.

46 Severe deficiency affects the young plants. Brown areas develop on leaves, particularly on the margins, and the surrounding tissue may become yellow (cv. Ostinata).

47 Yellowing of the leaves begins near the margins and, in summer crops, may progress rapidly from the older to the younger leaves (cv. Ostinata).
Plant analysis, 0.5% K.

47

48

Calcium deficiency

Growth is stunted and small brown (or blackish-brown) areas develop marginally and interveinally.

48 The leaf margins may be scorched, with the result that the expanding leaves become cupped. Growth becomes seriously restricted and small brown or blackish-brown areas develop interveinally; these areas occur on the youngest leaves when the deficiency becomes severe (cv. Ostinata).
Plant analysis, 0.23% Ca; deficiency, <1.0% Ca; normal range, 1–1.8% Ca.

49

Manganese deficiency

Growth may hardly be affected at first, but the plants become uniformly yellow-green in colour.

49 The interveinal areas of the outer leaves of young plants become pale brown, starting near the leaf tip (cv. Ostinata).
Plant analysis, 8 $\mu g\,g^{-1}$ Mn; deficiency, < 20 $\mu g\,g^{-1}$ Mn; normal range, 50–200 $\mu g\,g^{-1}$ Mn.

50 In mature plants, interveinal yellowing develops and spreads until only the veins remain green (cv. Ostinata).
Plant analysis, 9 $\mu g\,g^{-1}$ Mn.

50

51

Manganese toxicity

Growth and heart formation are restricted. Small brown areas develop near the leaf margins and parts of the outer leaves become yellow.

51 The margins of the oldest leaves of seedlings become pale yellow, beginning near the tips (cv. Ostinata).

52 Small brown areas, sometimes surrounded by yellowing, develop and coalesce near the margins of mature leaves; yellowing near the leaf tips may precede the appearance of the brown areas (cv. Ostinata).
Plant analysis, 1498 μg g^{-1} Mn; toxicity, >250–500 μg g^{-1} Mn, depending on cultivar.
The yield of susceptible cultivars may be depressed progressively as the Mn content of the leaves increases above 200 μg g^{-1} Mn.

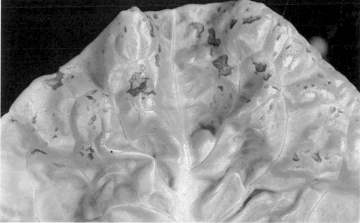

52

Copper deficiency

Growth is seriously restricted, the leaves are narrow and 'cupped' and the outer leaves may become yellow and die prematurely.

53 The narrow leaves become 'cupped' and tend to twist sideways to form an untidy whorl. The outer leaves usually become yellow and die prematurely (cv. Neptune).
Plant analysis, 1.8 μg g^{-1} Cu; deficiency, <2 μg g^{-1} Cu; normal range, 5–15 μg g^{-1} Cu.

53

54

Boron deficiency

Growth is restricted, the leaves are dark green and somewhat 'leathery', and in extreme cases the growing point dies.

54 Growth is very variable. With severe deficiency the outer leaves become yellow and die whilst the inner leaves become stiff, dark green in colour and tend to curl upwards. Less severely affected plants show some loss of colour and fail to heart.

55

55 The leaves of young plants may become strongly curled upwards (cv. Ostinata).
Plant analysis, 16 μg g^{-1} B; deficiency, <20 μg g^{-1} B; normal range, 30–60 μg g^{-1} B.

Boron toxicity

The margins of the leaves become scorched.

56 Small brown areas develop and coalesce on, or near, the margins of the older leaves (cv. Ostinata).
Plant analysis, 171 μg g^{-1} B (346 μg g^{-1} B in the outer leaves); toxicity, >80 μg g^{-1} B; normal range, 30–60 μg g^{-1} B.

57 The brown areas develop first near the tip of the leaf (cv. Ostinata).

56

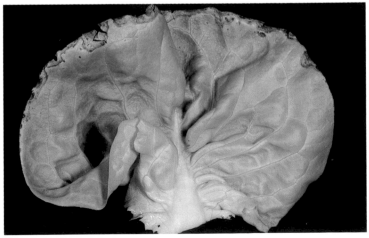

57

58

Molybdenum deficiency

Growth is severely restricted. The interveinal parts of the outer leaves become brown and the leaves die prematurely. Young seedlings are particularly vulnerable, and this disorder has occurred frequently on plants propagated in peat blocks. A severe deficiency can cause total crop failure.

58 Growth is seriously restricted (left) and the leaves become yellow-green in colour (cv. Ostinata).

59 The outer leaves become pale cream, and brown interveinal and marginal areas develop rapidly. The oldest outer leaves may remain green around the veins whilst all but the youngest growth becomes brown and withers (cv. Amanda). See also Volume 1, Plate 61.
Plant analysis, $0.10\ \mu g\ g^{-1}$ Mo; deficiency, $<0.12\ \mu g\ g^{-1}$ Mo; normal range, $>0.15\ \mu g\ g^{-1}$ Mo. Ranges based on the limited data available.

60 Severe deficiency occurs in slightly acidic peat (right; pH 5.0) whereas plants grown in peat limed to pH 6.7, though not supplied with Mo, grow normally (cv. Amanda).

59

60

Peppers

61

Nitrogen deficiency

Growth is seriously restricted and the foliage is yellow-green in colour.

61 Growth is restricted and the leaves become small and pale yellow-green in colour. The lower leaves become yellow and die prematurely. A normal plant is included for comparison (cv. Bellboy).
Leaf analysis of deficient plant, 1.2% N; deficiency, <2% N; normal range 3.5 – 5.5% N.

Phosphorus deficiency

Growth is restricted and, in contrast to N deficiency, the foliage remains dark green.

62 Growth is severely restricted by phosphorus deficiency (right), the leaves are small and dark green in colour and the margins tend to curl upwards and inwards. The lower leaves die prematurely (cv. Bellboy).
Leaf analysis of deficient plant, 0.1% P; deficiency <0.2% P; normal range, 0.3 – 0.8% P.

63 Leaves from deficient plants are dark green in colour but greatly reduced in size.

62

63

64

Sulphur deficiency

A rare disorder which could be confused with N deficiency, owing to the yellow-green colour of the leaves.

64 The leaves of S deficient plants (right) become pale yellow-green in colour (cf. N deficiency) and growth is restricted. The new leaves may become narrow and more pointed (cv. Bellboy).
Leaf analysis, 0.1% SO_4–S. Healthy leaves contained 0.37% SO_4–S (only limited analytical data available).

65

Potassium deficiency

Growth is restricted and small brown spots develop on the mature leaves.

65 Growth is restricted and small reddish-brown spots develop on the leaves; on young plants these usually spread from the leaf tips (cv. Bellboy).
Leaf analysis, 0.5% K; deficiency, <2.0% K; normal range, 3 – 6% K.

66 When the deficiency occurs on more mature plants, some interveinal and marginal yellowing of the leaves may also develop in addition to the spotting.

67 Severely affected leaves become almost covered with reddish-brown spots (cv. Bellboy).
Leaf analysis, 0.4% K.

66

67

68

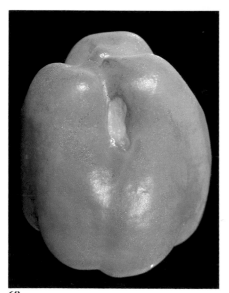

69

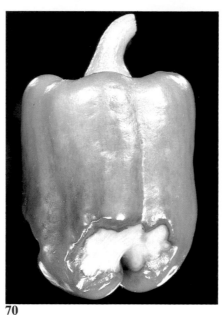

70

Calcium deficiency

The margins of the youngest leaves become yellow and pale brown sunken areas develop on the fruit, usually near the blossom end (similar to Ca deficiency in tomato).

68 Marginal yellowing of the youngest leaves, which usually develops first near the tips, spreads interveinally (cv. Bellboy).
Leaf analysis, 0.8% Ca; deficiency <1.0% Ca; normal range, 1.5 – 3.5% Ca.

69 Pale brown sunken areas develop near the blossom end of the fruit (cv. Bellboy).

70 With severe deficiency, the area of brown tissue becomes quite large and several parts of the same fruit may be affected. On less mature fruit, the whole of the distal end may be affected. Fruit analysis, 0.07% Ca; deficiency, <0.08% Ca; normal range, 0.1 – 0.2% Ca, but not always a reliable guide to the incidence of blossom-end rot.

71

Magnesium deficiency

The mature leaves develop interveinal yellowing.

71 An interveinal yellow-green chlorosis develops on mature leaves, though some dark green areas may remain beside the main veins (cv. Bellboy)
Leaf analysis, 0.12% Mg: deficiency, <0.30% Mg; normal range, 0.35 – 0.80% Mg.

72

73

Iron deficiency

The youngest leaves become yellow and interveinal yellowing develops near the base of the older leaves.

72 The youngest leaves are yellow and may become almost white. The tips remain green at first whilst the yellowing spreads from the base of the leaf (cv. New Ace. See also Volume 1, Plate 26).
Leaf analysis, 66 μg g^{-1} Fe; normal range, 80 – 200 μg g^{-1} Fe. The total Fe content is not a reliable indicator of Fe status.

73 On mature plants, an interveinal yellowing develops near the base of the leaves (cv. Bellboy).

74

Manganese deficiency

The young leaves become bright yellow-green and develop dark brown interveinal areas. Small, somewhat diffuse yellow areas develop on the mature leaves; these may become brown later.

74 The young leaves become bright yellow-green in colour and small dark brown areas develop interveinally usually beginning near the tip of the leaf. When severely affected, the leaves wither and drop off (cv. New Ace).
Leaf analysis, 5 μg g^{-1} Mn; deficiency, <20 μg g^{-1} Mn; normal range, 100 – 300 μg g^{-1} Mn.

75

Manganese toxicity

The oldest leaves become yellow-orange and die prematurely.

75 Yellow-orange areas develop on the lower leaves; these gradually spread across the entire surface and the leaves die prematurely (cv. Bellboy) Leaf analysis, 1003 μg g^{-1} Mn.

76

Copper deficiency

Growth is restricted, but the leaves remain dark green.

76 Growth of the deficient plant (right) is restricted and the leaf size is reduced. The leaf margins tend to curl upwards, but no other characteristic foliar symptoms have been identified (cv. Bellboy).
Leaf analysis, 7.0 μg g^{-1} and 1.7 μg g^{-1} Cu (left and right respectively); deficiency, <4 μg g^{-1} Cu; normal range, 6 – 20 μg g^{-1} Cu.

77

Zinc deficiency

Small purple areas, which eventually become brown, develop on most of the leaves.

77 The dark green leaves develop small purple areas which are scattered at random interveinally. These areas become pale brown as they enlarge (cv. New Ace; see also Volume 1, Plate 47.)
Leaf analysis, 6 μg g^{-1} Zn; deficiency, <25 μg g^{-1} Zn; normal range, 40 – 100 μg g^{-1} Zn.

78

Boron deficiency

Yellowing on the tips of the mature leaves gradually spreads around the margins, and the main veins become reddish-brown.

78 The new leaves of young plants become distorted when the supply of B is inadequate (cv. Bellboy).

79 The tips of the mature leaves become yellow and the foliage is somewhat brittle. (cv. Bellboy). Leaf analysis, 7 $\mu g\ g^{-1}$ B; deficiency, <20 $\mu g\ g^{-1}$ B; normal range, 30 – 90 $\mu g\ g^{-1}$ B.

80 The yellowing spreads around the margin from the leaf tips (cv. Bellboy).

81 Reddish-brown deposits in the veins are clearly visible when viewed by transmitted light.

79

80

81

82

Molybdenum deficiency

The foliage becomes yellow-green in colour and growth is somewhat restricted.

82 The leaves of seedlings become yellow and growth is restricted (cv. Bellboy). This deficiency occurs most commonly on acidic (pH < 5.0) substrates.

83 The younger leaves of mature plants become pale yellow-green in colour (cv. Bellboy).

83

132

Tomato

84

85

86

Nitrogen deficiency

Leaves are pale green and growth is restricted. Lower leaves become yellow and may develop areas of purple pigmentation before dying prematurely. New growth has a spiky appearance.

84 The leaves of N deficient plants (right) are pale green, their size is reduced and expansion of the basal leaflets is restricted (cv. Craigella).
Leaf analysis, 4.5% and 2.4% N in the normal and deficient leaves respectively; deficiency, <2.5% N; normal range, 3.5 - 5.0% N.

Phosphorus deficiency

Growth is seriously restricted, the undersides of the younger leaves usually become purple, and the older leaves die prematurely.

85 The youngest leaves are small and medium or dark grey-green in colour. Purpling develops on the petioles and undersides of the leaves, though purpling of the petioles also occurs on normal leaves in some cultivars (cv. Sarina).

86 Development of purple pigmentation on the undersides of the younger leaves (also on the upper surfaces when the deficiency is severe) is characteristic of the deficiency, though rather similar symptoms may occur on plants grown with insufficient heat (cv. Sarina).

87

88

Phosphorus deficiency continued

87 Small brown areas develop interveinally on the lower leaves (cv. Sonato).
Leaf analysis, 0.07% P; deficiency, <0.2% P; normal range, 0.35 - 0.75% P.

88 In some cultivars the brown areas are larger, their distribution apparently being random (cv. J 348).

89

90

Sulphur deficiency

A rare disorder which, due to the pale colour of the leaves, could be confused with N deficiency.

89 The leaves become yellow-green with purple veins and petioles. The petioles fail to elongate normally, giving the leaves a compressed appearance (cv. Sonato).
Leaf analysis, 0.10% SO_4–S; leaves from healthy plants contained 0.29% SO_4–S.

90 A pale green, S deficient leaflet with a normal leaflet for comparison; note the pink/purple colouration in the veins. The deficient leaflets tend to be puckered around the main veins.

91

92

93

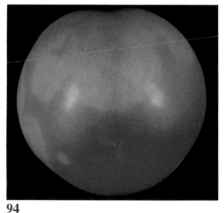

94

Potassium deficiency

The margins of the upper leaves become yellow, later brown, and the fruit ripen unevenly.

91 The margins of the young, fully expanded leaves become yellow and scorch (pale brown) rapidly in hot weather (cv. Selsey Cross).

Leaf analysis, 2.2% K; deficiency, <2.5% K; severe deficiency, <1.0% K; normal range, 3.5 - 6.3% K.

92 The marginal yellowing spreads interveinally as the deficiency progresses, and small, pale brown areas develop; these are usually, though not exclusively, interveinal (cv. Sonato).

Leaf analysis, 0.5% K.

93 Some yellowing occurs near the margins of the leaves on out-door plants, but marginal browning is the predominant feature (cv. Sonatine).

Leaf analysis: 1.35% K; leaves from healthy plants contained 3.1% K.

94 When grown with insufficient K, ripening of parts of the fruit wall is delayed, resulting in green (later yellow) areas on an otherwise red fruit. The size of the abnormal areas depends on the severity of the deficiency (cv. Grenadier). Potassium deficiency is not, of course, the only factor that causes uneven ripening of tomato fruit.

95

96

97

98

Calcium deficiency

Lack of available Ca results in scorching of the new growth, death of the growing point and blossom-end rot of the fruit. A temporary shortage of Ca may cause yellowing of the tips of rapidly expanding leaves (see under 'Physiological disorders', Plates 132 and 133).

95 The leaves of seedlings become distorted and develop yellow, interveinal areas; the growing point soon dies.

96 On mature plants, the margins of the youngest leaves become brown and some interveinal areas become yellow. The growing point dies and flower buds fail to develop (cv. Sonato).
Leaf analysis, 0.2% Ca; deficiency, <1.0% Ca; normal range; 2.0 – 4.0% Ca.

97 The margins become brown at the base of the leaflets (cv. Sonato).
Leaf analysis, 0.18% Ca.

98 In some cases, the leaflets become yellow at the tip before any brown areas develop (cv. Sonatine).
Leaf analysis, 0.08% Ca (laminae only).
See Plate 132 for similar symptoms in response to a prolonged period of humid conditions.

99

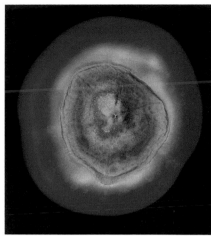

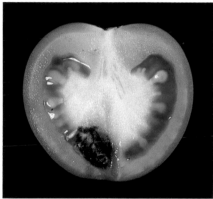

100

101

102

99 Blossom-end rot develops on fruit receiving insufficient Ca. Affected fruit generally contain less than 0.08% Ca whereas healthy fruit frequently contain 0.12 - 0.25% Ca.

100 Fruit with blossom-end rot ripen prematurely and are inedible.

101 Less severe Ca stress may result in the formation of sunken greyish-brown spots near the stylar scar and blackening of part of the placenta ('internal blossom-end rot'; cv. Marathon).

102 When blossom-end rot develops on very young fruit, a much larger proportion of the fruit is affected. Two degrees of severity are shown.

103

Magnesium deficiency

Interveinal yellowing develops on the middle and lower leaves.

103 The deficiency symptoms first appear on the middle and lower leaves and gradually progress up the plant.

104 Yellow interveinal areas develop whilst the main veins, and frequently the margins of the leaves, remain green (cv. Sonato).
Leaf analysis, 0.16%Mg; deficiency, <0.3% Mg; normal range 0.35 - 0.80% Mg.

105 The lower leaves of severely deficient plants become yellow-orange, and areas of purple pigmentation may develop (cv. Nemato).
Leaf analysis, 0.14% Mg.

104

105

106

Iron deficiency

Leaves in the head of the plant became pale yellow, whilst the young mature leaves develop a mottled appearance.

106 The young leaves become pale green and develop interveinal yellowing which spreads from the base towards the tips of the leaflets (cv. Selsey Cross). See also Volume 1, Plate 24.

Leaf analysis, 54 $\mu g \, g^{-1}$ Fe; deficiency, <60 $\mu g \, g^{-1}$ Fe; normal range, 80 - 200 $\mu g \, g^{-1}$ Fe. The total Fe content is not always a reliable indicator of Fe status; see p62.

107 When the deficiency is not corrected, brown areas usually develop along and around the main veins, particularly near the base of the leaflets, and on the petioles. The youngest leaves may become almost white (cv. Selsey Cross).

108 Symptoms of Fe deficiency develop in very young plants when the roots are poorly aerated. The seedling shown was growing in a rockwool cube that became waterlogged.

107

108

109

Manganese deficiency

Insufficient Mn results in yellowing of the younger leaves. Persistence of a uniformly dark green veinal network in the yellow leaves is characteristic of this deficiency and distinguishes it from Fe deficiency.

109 The interveinal chlorosis usually appears first on the smallest leaflets (cv. Sonato.) Leaf analysis, $18\mu g\,g^{-1}$ Mn; deficiency, $<25\,\mu g\,g^{-1}$ Mn; normal range, $100 - 300\,\mu g\,g^{-1}$ Mn.

110 As the deficiency develops, the interveinal areas become more uniformly yellow (cv. Sonato); compare with Fe deficiency where the yellowing begins at the base of the leaflets. Leaf analysis, $8\mu g\,g^{-1}$ Mn.

111 The veinal network remains dark green. Eventually small brown spots develop in the yellow, interveinal areas.

110

111

Manganese toxicity

Excessive amounts of available Mn are released when some soils are partially steam-sterilised. Plants grown in such soils accumulate Mn in the lower leaves; black deposits develop in the tissue along the lines of the veins and these leaves die prematurely. The problem may be aggravated by overwatering and relieved by heavy applications of lime and P.

112 As the black deposit around the veins increases, yellowing develops in the surrounding tissue and gradually spreads across the leaf (cv. Sonato). Leaf analysis, $4000\,\mu g\,g^{-1}$ Mn. Mild toxicity symptoms may occur on leaves containing 1000-$1500\,\mu g\,g^{-1}$ Mn and severe symptoms on those with >2500 $\mu g\,g^{-1}$ Mn.

112

113

114

113 With severe toxicity, the black deposit in the veins spreads into the surrounding tissue. Affected leaves die rapidly (cv. Minicraigella).
Leaf analysis, 8166 µg g^{-1} Mn.

114 In cases of severe Mn toxicity, purple spots develop on the stems and petioles and soon become pale brown in colour; this condition is here shown in five short lengths of tomato stem. The calyces of the fruit (not illustrated) also became scorched at the tips.

115

116

Copper deficiency

Wilting of the youngest leaves of plants which are adequately supplied with water is characteristic of Cu deficiency. This rare disorder is usually associated with highly organic substrates such as peat.

115 The youngest leaves become dark green and, after frequent wilting, the margins become scorched. These symptoms somewhat resemble those induced by high salinity. When the deficiency becomes as severe as this, the flower buds fail to develop (cv. Sonato). See also Volume 1, Plate 39.

116 The margins of the mature leaves tend to curl upwards and inwards (cv. Sonato).
Leaf analysis, 2.5 µg g^{-1} Cu; deficiency, <4 µg g^{-1} Cu; normal range, 7 - 20 µg g^{-1} Cu.

117

118

Copper deficiency continued

117 Pale spots develop, firstly near the leaf tips and usually on or near a vein (cv. Sonato). Leaf analysis, 1.7 $\mu g\,g^{-1}$ Cu.

118 The upper part of a tomato plant severely affected by Cu deficiency showing a brown leaf scorch in the head (cf. Plate 115) and pale spotting of the expanded leaves below. The spots tend to coalesce, particularly at the leaf tips.

119

Zinc deficiency

119 The leaflets become curled downwards and inwards. The main veins remain green but the interveinal areas become yellow. Brown areas develop near the midrib and main veins and on the petiolules (cv. Sonatine). Leaf analysis, 8 $\mu g\,g^{-1}$ Zn; deficiency, <20 $\mu g\,g^{-1}$ Zn; normal range, 35 - 100 $\mu g\,g^{-1}$ Zn.

Zinc toxicity

120 Growth is restricted and becomes spindly.

121 Interveinal yellowing develops on the leaflets whilst the petioles and undersides of the veins become purple. The small leaflets remain only partly unfolded (cv. Sonatine). Leaf analysis, 280 $\mu g\,g^{-1}$ Zn

121

120

122

Boron deficiency

Yellowing of the tips of the lower leaves is usually the first symptom of B deficiency to be noticed on either young or mature plants. The yellowing spreads around the margins of the leaves and progresses up the plant. Brittleness of the foliage is characteristic. Severe B deficiency results in death of the growing point.

122 Boron deficiency frequently occurs towards the end of the propagation stage on plants not receiving B in the liquid feed.

123 The tips of the lower leaves of young plants become yellow and the chlorosis gradually spreads across the leaves. Leaf analysis, $8 \, \mu g \, g^{-1}$ B; deficiency, $<25 \, \mu g \, g^{-1}$ B; normal range, $30 - 80 \, \mu g \, g^{-1}$ B.

124 On mature plants, purple areas may develop on the tips of leaves which become yellow.

125 Yellowing and browning of the leaf tips is frequently associated with a small break across the main vein of the leaflet (centre and right), which results from the characteristic brittleness of the foliage.

123

124

125

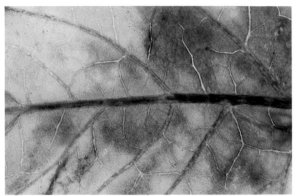

126

Boron deficiency continued

126 As the deficiency becomes more severe, purplish-brown deposits form in the veins; these are best viewed by transmitted light i.e. through the leaf from the underside.

127 Corky areas develop near the calyx or on the shoulders of the fruit on plants seriously deficient in B.

127

Boron toxicity

Excessive uptake of B causes a marginal scorch on the foliage and may result in death of the plant (borax has sometimes been used as a weed killer).

128 Small brown spots, which later enlarge and coalesce, develop along the margins of the leaflets and the tips of the calyces on the fruit become scorched. The marginal scorch shown here developed on plants grown in a recirculating nutrient solution containing 2 mg l^{-1} B (cv. Sonato).
Typical leaf analysis, 220 μg g^{-1} B; toxicity $>150 \text{ μg g}^{-1}$ B

129 Application of B at high concentrations results in severe toxicity, shown here by death of extensive areas of interveinal tissue. The lower leaves become yellow and die rapidly as the symptoms spread up the plant (cv. Sonato).
Typical leaf analysis, $350 - 650 \text{ μg g}^{-1}$ B.

128

129

130

Molybdenum deficiency

The foliage is yellow-green and a pale yellow margin develops on the older leaves.

130 A pale, Mo deficient tomato plant on the right, with a normal plant on the left for comparison. Young plants grown under acid conditions (pH <5.0) are particularly susceptible to this deficiency.

131 The older leaves develop a pale yellow margin which spreads interveinally. Later, the margins become almost white or very pale brown (cv. Selsey Cross). See also Volume 1, Plate 60. Leaf analysis, $0.1\ \mu g\,g^{-1}$ Mo; deficiency, $<0.3\ \mu g\,g^{-1}$ Mo; normal range, $>0.4\ \mu g\,g^{-1}$ Mo (only limited data available).

131

132

Physiological disorders

132 Yellowing of the tips, and later of the margins, of the rapidly expanding leaves (3rd to 5th below the growing point) may develop after the rate of transpiration has been greatly depressed for several days, e.g. by dull, humid weather (cv. Nemato). When these conditions persist, pale brown areas develop and these may be colonised rapidly by *Botrytis cinerea*. This disorder, believed to be associated with restricted movement of Ca in the transpiration stream, may be prevented or corrected by spraying with a solution of 0.2 - 0.3% calcium nitrate.

133 When excessive humidity is maintained, the tips of expanding leaves become yellow and fail to develop normally (cv. Sonatine).

133

134

135

136

Physiological disorders continued

134 At very high humidity the tips of leaflets on the youngest leaves bend inwards and become black (cv. Sonatine).

135 In solution culture, roots are sometimes damaged when the plants or roots are moved. Small, pale brown spots develop in the interveinal areas of the young expanded leaves. The severity of the spotting increases with the extent of the root damage, usually affecting 3–5 leaves. Similar symptoms occur after root pruning (cv. Marathon).

136 When the roots of plants grown in solution culture are allowed to dry out, pale brown areas develop between the veins of the lower leaves. The symptoms spread rapidly up the plant (cv. Marathon).

Carnation

137

138

139

Nitrogen deficiency

Growth is somewhat restricted and the foliage is generally paler than normal in colour.

137 A pale green, deficient leaf (left), together with a normal leaf for comparison (cv. Pink Sim). Leaf analysis, 0.4 and 2.5% N in the deficient and normal leaves respectively; deficiency, <2.0% N; normal range, 2.5 – 3.8% N.

Phosphorus deficiency

Growth is seriously restricted. The foliage is dark blue-green in colour (cf. N deficiency), but the oldest leaves become pale brown and wither.

138 Growth is greatly restricted and the lower leaves die prematurely. The upper leaves are narrow and blue-green in colour (cv. White Sim).

139 The lower leaves develop pale brown areas which spread rapidly over the entire leaf surface (cv. White Sim). Leaf analysis, 0.03% P; deficiency, <0.15% P; normal range, 0.20 – 0.40% P.

140

140 At very low levels of P the leaves of the young vegetative shoots are blue-green and exceptionally narrow. The tip scorch spreads slowly towards the base of the leaves (cv. William Sim).

141 After a bloom has been harvested, one or more pairs of leaves below the cut die rapidly as new shoots develop (cv. White Sim).

141

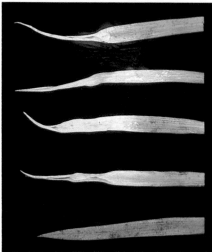

142

Potassium deficiency

Tip burn occurs on the lower and middle leaves, and pale, sunken spots develop in the middle and upper foliage. This leaf spotting is particularly characteristic of K deficiency in the commercially important Sim cultivars.

142 Pale brown areas develop at the tip and progress towards the base of the leaves (cv. Saugus White).
Leaf analysis, 4.1 and 1.7% K in the normal (bottom) and deficient leaves respectively; deficiency, <2.0% K; normal range, 2.5 – 4.5% K.

143 Small, sunken white spots develop in the middle and upper leaves (cv. White Sim).
Leaf analysis, 4.3 and 1.9% K in the normal and deficient leaves respectively.

144 The white spots, usually without tip burn, may also affect the small leaves just below the bloom.

143

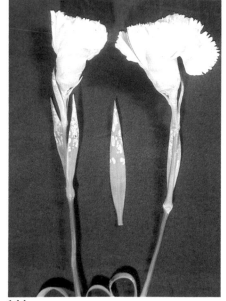

144

145

Calcium deficiency

The margins of the expanding leaves curl upwards and inwards near the tips, which rapidly become scorched. These symptoms sometimes occur when a newly planted crop begins to make rapid growth and usually reflect the low Ca status of the stock plants rather than a deficiency of Ca in the new substrate.

145 The margins of expanding leaves curl upwards and inwards and the restriction formed where the curled area joins the rest of the leaf blade (right) is characteristic. The tips of the affected leaves rapidly become scorched (pale brown). Leaf analysis, 0.50% Ca; deficiency, <1.0% Ca; normal range, 1.2 – 1.8% Ca.

146 A stock plant showing symptoms of Ca deficiency. Leaf analysis, 0.81% Ca in leaves with scorched tips.

146

147

Iron deficiency

The younger leaves become yellow and may later become almost white.

147 A young plant showing yellowing of the leaves on the new shoots (cv. Pink Sim).

148 The lateral regions of the younger leaves turn yellow and may later become almost white. Tissue adjacent to the mid-rib remains green at first. A normal leaf is included below for comparison (cv. Pink Sim). Leaf analysis, 100 and 34 μg g^{-1} Fe in the normal and deficient leaves respectively; deficiency, <60 μg g^{-1} Fe; normal range, 75 – 125 μg g^{-1} Fe.

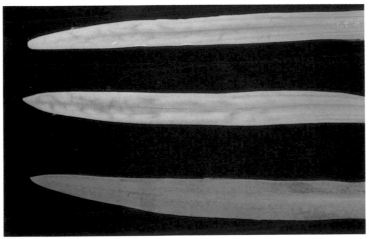

148

149

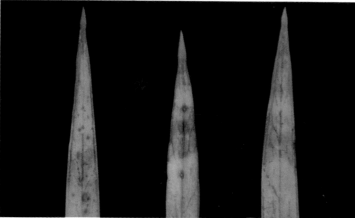

150

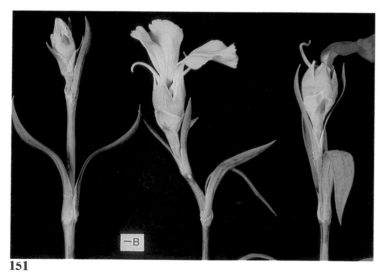

151

Boron deficiency

The tips of some of the upper leaves become purple, the incidence of calyx splitting is markedly increased and some of the blooms may be distorted or fail to develop. This is the most important micronutrient deficiency of carnations encountered commercially.

149 Three B deficient leaves showing purpling near their tips, together with a normal leaf (below) for comparison. The tips slowly wither, leaving a narrow purple-red band between the green and the dead tissue. In severe instances of deficiency, as in this Plate, a zone of pale yellow or cream coloured tissue also develops along the mid-rib, together with small scattered patches of purple colour (cv. Pink Sim).
Leaf analysis, 9 and 129 μg g^{-1} B in the deficient and normal leaves respectively; deficiency, <25 μg g^{-1} B; normal range, 30 – 80 μg g^{-1} B.

150 The band of purple pigmentation which forms initially at the leaf tip (cf. Plate 149) gradually progresses towards the base of the leaf as the tip dies back.

151 The blooms of B deficient carnations may be distorted and, in extreme cases, very few petals develop (cvs White Sim and William Sim). The incidence of calyx splitting is also greatly increased (see Table 25).

152

153

154

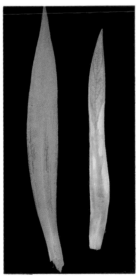

155

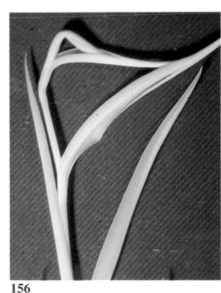

156

Boron deficiency continued

152 With a severe deficiency, some of the buds fail to develop, turn pale brown and wither. Once the growing point has died, an excessive number of axillary growths may develop (cv. Pink Sim).
Leaf analysis, 8 μg g^{-1} B.

153 An example of the multiple branching that may result from the loss of apical dominance associated with death of the flower bud.

Molybdenum deficiency

The foliage is pale green and the leaves become scorched.

154 The leaves become pale green in colour and the margins near the middle and base of young mature and expanding leaves become scorched (cv. Pink Sim).

155 The very pale brown areas that develop on the margins near the middle and base of the deficient leaves (right) gradually spread both along the margins and inwards towards the main vein (cv. Pink Sim).
Leaf analysis, 1.3 and 0.1 μg g^{-1} Mo in the normal and deficient leaves respectively; deficiency, <0.3 μg g^{-1} Mo; normal range, >0.5 μg g^{-1} Mo (only limited data available).

Physiological disorders

In winter, growth and flower production are limited by poor light; stem strength is reduced, and the separation of new leaves is inhibited at low levels of N.

156 The new leaves on N deficient plants are tightly rolled and frequently fail to separate at the tips during winter. They develop normally, however, when the light improves. This condition is sometimes referred to as 'curly tip'.

Chrysanthemum

157

Nitrogen deficiency

Growth is restricted and the leaves become yellow-green. Root development is usually stimulated when the deficiency is not too severe.

157 Part of a factorial nutritional study, showing pale green N deficient plants in the foreground (plot 27). The taller, dark green plants in plot 26 had received a fully adequate supply of N (cv. Hurricane).
Leaf analysis, 4.0 and 1.8% N in leaves of the normal and deficient plants respectively.

158 The four plants show responses in the growth, leaf and flower size and leaf colour to increasing applicaions of N. A severe deficiency (left) delays flowering and reduces the number and size of the blooms (cv. Hurricane).
Leaf analysis, 1.9, 2.1, 3.6 and 5.5% N respectively from left to right; deficiency, <2.5% N; normal range 3.5–5.5% N.

158

159

Nitrogen excess

Heavy applications of organic fertilisers may result in root damage, particularly after partial sterilisation of the substrate. The distribution of affected plants is rather irregular, those with chlorotic upper leaves normally being surrounded by very dark green plants that are somewhat stunted.

159 Uneven distribution of ureaformaldehyde (38% N), applied to a bed of steamed peat before planting, resulted in high concentrations of NH_4-N in some parts of the bed (cv. Hurricane; four weeks after planting in early April).

160 A single plant showing yellowing of the upper leaves due to excess N (cv. Hurricane). Before planting, ureaformaldehyde (38% N) had been applied to the soil at the rate of $100 \, \text{g m}^{-2}$.

160

161

Phosphorus deficiency

Growth is seriously restricted. The upper leaves become dull grey-green in colour and the oldest leaves become yellow or orange and die prematurely.

161 The lower leaves of young plants become yellow and die. The upper leaves become grey-green and remain closely spaced due to the slow rate of stem elongation (cv. Hurricane). Leaf analysis, 0.1% P; deficiency, <0.2% P; normal range, 0.30–0.80% P.

162

163

164

165

Phosphorus deficiency continued

162 Growth is seriously restricted, fewer buds are formed, flowering is delayed and the lower leaves die prematurely (cv. Hurricane).

163 Leaves from a mature plant showing marginal browning on a lower leaf (bottom of picture) and yellow or orange-brown margins and areas developing on middle (centre) and upper leaves (cv. Hurricane).
Leaf analysis, 0.09% P.

Sulphur deficiency

The leaves become yellow-green in colour and growth is restricted. This rare disorder could easily be confused with N deficiency.

164 The leaves of S deficient plants become pale yellow-green (later yellow) in colour, though the leaf tips remain green at first; compare with the normal leaf below (cv. Fandango).
Leaf analysis, 0.07% and 0.19% SO_4-S in the deficient and normal leaves respectively.

Potassium deficiency

A yellow or brown margin develops on the leaves but, depending on the severity, growth may not be greatly affected.

165 Browning of the margins of the lower leaves spreads both inwards across the leaf surface and upwards on the plant (cv. Hurricane).

166

167

166 When the deficiency becomes severe, the margins of the upper leaves become brown, the number and the size of the blooms is reduced and flowering may be slightly delayed (cv. Hurricane).
Leaf analysis, 1.2% K.

167 In some cultivars, browning may spread from the tip of the leaf towards the base rather than inwards from the entire margin (basal leaves; cv. Fred Shoesmith).
Leaf analysis, 1.6% K; deficiency, <2.0% K; normal range, 3.5–6.0% K.

168

169

170

Calcium deficiency

Growth is restricted, the leaf tips may become scorched and brown areas may develop on the upper leaves. When the deficiency becomes severe, the growing point or the flower buds may die.

168 The plants are stunted and the tips of the expanding leaves become scorched; these leaves remain shortened and distorted (cv. Hurricane).
Leaf analysis, 0.46% Ca; deficiency, <0.8% Ca; normal range, 1.0–2.5% Ca.

169 When the deficiency becomes severe after bud formation, the buds become brown and fail to develop further (cv. Hurricane).

170 Dull brown areas may develop interveinally and near the margins of the expanding leaves (cv. Fandango).
Leaf analysis, 0.29% Ca.

171

Magnesium deficiency

Interveinal yellowing develops on the older leaves but growth may not be greatly affected.

171 The interveinal areas of the older leaves become yellow, but the main veins remain green (cv. Flaxen Princess Anne).
Leaf analysis, 0.05% Mg; deficiency, <0.1% Mg; normal range, 0.25–1.0% Mg.

Many cultivars appear to tolerate a low Mg status. For example, the leaves of cv. Hurricane were fully developed and dark green although they contained only 0.12% Mg.

172

Iron deficiency

The upper leaves become yellow (almost white when the deficiency is severe) and brown areas may develop and enlarge. Fewer buds are formed and flowering may be delayed.

172 The young leaves of newly-established cuttings become yellow when inadequately supplied with Fe (cv. Hurricane). Leaf analysis, 30 μg g^{-1} Fe.

173 The young plants frequently recover from an early, mild Fe deficiency without further treatment, and the new leaves become progressively greener; the veins, however, may be much paler than those in normal leaves. Compare an older deficient leaf (top of picture) with a younger leaf in the centre; the younger leaf is greener, indicating partial recovery from the initial deficiency. A normal green leaf is shown below (cv. Pollyanne). Leaf analysis, 53, 60 and 91 μg g^{-1} Fe respectively from top to bottom.

174 When Fe deficiency affects a more mature plant the upper leaves become yellow or pale cream and brown areas may develop and enlarge (cv. Fandango).
Leaf analysis, 21 μg g^{-1} Fe; deficiency, <60 μg g^{-1} Fe; normal range, 80–200 μg g^{-1} Fe.

173

174

175

176

Manganese deficiency

The plants become uniformly pale green and interveinal yellowing develops on the mature leaves. The interveinal pattern formed varies with cultivar. Growth may be restricted, fewer buds formed, and flowering delayed.

175 The Mn deficient plants (right) become yellow-green and growth is somewhat restricted (cv. Fandango).
Leaf analysis, 135 and 8 μg g^{-1} Mn in leaves from the normal and deficient plants respectively.

176 The leaves become yellow except for the veinal network which remains green (centre and right; cv. Hurricane). The veinal pattern is modified in some cultivars because bands of tissue beside the veins also remain green.
Leaf analysis, 14 and 100 μg g^{-1} Mn respectively in the deficient and normal leaves; deficiency, <20 μg g^{-1} Mn; normal range, 50–300 μg g^{-1} Mn.

177 Flowering is delayed by the deficiency (right) and the number and size of the blooms are reduced (cv. Hurricane).

177

178

Manganese toxicity

The margins of the upper leaves and of the small leaves on the axillary growths become yellow. Black spots may develop on the lower leaves, which then become yellow and die prematurely.

178 On young plants, excess Mn causes yellowing of the margins of the upper leaves, though the small leaves on the axillary growths are usually affected first. Later, the younger leaves may become almost entirely yellow (cv. Fandango).
Leaf analysis, 862 $\mu g\,g^{-1}$ Mn; toxicity, >800 $\mu g\,g^{-1}$ Mn, with values up to 2600 $\mu g\,g^{-1}$ Mn recorded.

179 The top of a young plant (left) affected by excess Mn (cv. Hurricane).

180 Excess Mn accumulates in the lower leaves of mature plants; the veins become blackened, dark spots form near the veins, and the leaves turn yellow (left and centre) and die prematurely (cv. Polaris).
Leaf analysis, 935 $\mu g\,g^{-1}$ Mn.

179

180

181

182

183

184

Copper deficiency

Pale green leaves and delayed flowering are characteristic of this deficiency. Growth is seriously restricted and flowering may be suppressed by a severe deficiency, whereas a mild deficiency results in elongation, and sometimes excessive branching, of the lateral shoots, which carry several leaves and enlarged bracts below each bud.

181 A severe deficiency (right) inhibits growth of the lateral shoots, prevents flowering and results in premature death of the lower leaves (cv. Heyday). Leaf analysis, 2.5 μg g^{-1} Cu; deficiency, <4.0 μg g^{-1} Cu; normal range, 7–20 μg g^{-1} Cu.

182 When flowering is seriously delayed, enlarged bracts continue to form on the lateral shoots until buds are initiated (left). When flowering is completely suppressed (right) the axillary shoots bear normal leaves and develop according to the severity of the deficiency (cv. Hurricane). Leaf analysis, 2 μg g^{-1} Cu. The growth of plants with moderate or severe deficiency appears to be limited, thereby maintaining a minimum concentration of 1.5–2.0 μg g^{-1} Cu in the leaves.

183 Three axillary shoots showing the branching that frequently occurs when flowering is suppressed (left) or seriously delayed (right). Note the comparatively short pedicel of a bloom from the corresponding position on a normal plant (centre); such blooms carry, at the most, only one or two very small bracts on the pedicel (cv. Hurricane).

184 The margins and interveinal areas of the leaves of deficient plants (right) become pale yellow-green or yellow, but a band of green remains around the main veins (cv. Pollyanne). Leaf analysis, 9.2 and 3.4 μg g^{-1} Cu respectively in the normal and deficient leaves.

185

186

187

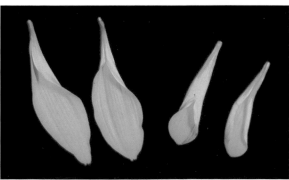

188

Copper deficiency continued

185 Chrysanthemums treated with growth retardants, as is normal in the commercial production of pot-grown plants, are less seriously affected by the deficiency (right). The leaves become pale green or yellow-green, flowering is slightly delayed (<10 days) and bloom diameter may be reduced (cv. Yellow Bonnie Jean, treated with two applications of 0.5% daminozide).
Leaf analysis, 14.0 and 2.9 μg g^{-1} Cu respectively in the normal and deficient leaves.

Excess copper

High levels of Cu in the substrate result in yellowing of the upper leaves of established cuttings.

186 Rapidly expanding leaves become yellow-green or yellow but the margins and tips may remain green. Yellowing around the veins is sometimes particularly marked (cv. Hurricane).

Boron deficiency

The leaves remain dark green and become very brittle. The flower petals become 'quilled' i.e. fail to unroll properly or to reflex fully, and bruise easily. When the deficiency is very severe, growth is restricted, purple colouration may develop on the leaf tips in some cultivars and the flowers are small and distorted.

187 The petals have a spiky appearance (left) compared with those of a normal bloom (right), and bruise easily (cv. Pollyanne).
Leaf analysis, 15 μg g^{-1} B; deficiency, <20 μg g^{-1} B; normal range, 30–80 μg g^{-1} B.

188 The affected petals (right) fail to unroll properly, remaining tubular or 'quilled' (cv. Bright Golden Anne).

189

190

Boron toxicity

Scorching (brown) of the tips and margins of the leaves. Growth may be restricted.

189 Excess B scorches the tips and then the margins of the lower leaves (cv. Pollyanne).
Leaf analysis, 145 μg g^{-1} B; toxicity, >100 μg g^{-1} B.
In some cultivars e.g. Hurricane, the scorched areas are a deeper brown colour.

190 A detached leaf showing the narrow brown margin of scorched tissue caused by excess B. The browning is most extensive around the tips of lobes of the leaves, which is where the symptoms first appear.

191 When the toxicity becomes severe, the margins of the upper leaves also become scorched and the browning spread inwards from the margin (cv. Hurricane).

192

Molybdenum deficiency

The leaves are pale green and yellowing develops on the mature young leaves. This deficiency is rare, as cuttings produced commercially usually contain sufficient Mo to sustain the mature plants.

192 The leaves of deficient plants (left) are pale green, and become yellow, firstly around the margins and later interveinally (cv. Fandango).
Leaf analysis, 0.10 and 1.1 μg g^{-1} Mo respectively in the deficient and normal leaves; deficiency, <0.3 μg g^{-1} Mo; normal range, >0.5 μg g^{-1} Mo (only limited data available).

191

193

Physiological disorders

Plants grown at a low level of N are particularly susceptible to damage by nicotine fumigation.

193 A yellow margin develops around many of the leaves shortly after fumigation (cv. Hurricane).

Poinsettia

194

195

Nitrogen deficiency

Growth is restricted and the leaves become yellow.

194 N deficient plants (right) are pale yellow-green in colour and growth is restricted (cv. Annette Hegg).

195 The leaves are uniformly yellow-green in colour and much reduced in size (right). The pink colouration in the veinal network is particularly intense at first, but fades as the deficiency becomes more severe (cv. Annette Hegg). Leaf analysis, 4.3 and 1.1% N in the normal and deficient leaves respectively; deficiency <2.5% N; normal range, 4.0–5.5% N.

196

Phosphorus deficiency

Growth is seriously restricted and the lower leaves become yellow and die prematurely.

196 A P deficient plant (right), with a normal plant for comparison (cv. Annette Hegg).

197

198

Phosphorus deficiency continued

197 The first symptom of deficiency is a loss of colour in the lower leaves (centre and right; cv. Annette Hegg).

198 With severe deficiency, the leaf size is greatly reduced (centre and right) and the tips of the yellow leaves become brown before they die prematurely (cv. Annette Hegg).
Leaf analysis, 0.49 and 0.04% P in the normal and deficient leaves respectively; deficiency, <0.2% P; normal range, 0.35–0.75% P.

199

Potassium deficiency

The margins and interveinal areas of the mature leaves may become yellow before pale reddish-brown spots develop; these coalesce, and the margins may rapidly become scorched.

199 Yellowing of the margins of the deficient leaves (right) spreads interveinally (cv. Annette Hegg).

200 With more severe deficiency (right), pale reddish-brown spots develop on mature leaves both interveinally and near the margins (cv. Annette Hegg).
Leaf analysis, 4.5 and 0.6% K in the normal and deficient leaves respectively; deficiency, <1.5% K; normal range, 2.5–5.0% K.

200

201

201 The reddish-brown spots coalesce and the margins may rapidly become scorched (right). Growth may not be affected seriously but the plants are disfigured (cv. Annette Hegg). Leaf analysis, 0.2 and 3.9% K in the deficient and normal leaves respectively.

202

Magnesium deficiency

Interveinal yellowing and scorching develops on the lower leaves.

202 The leaves of Mg deficient plants (left) become pale yellow-green except for the dark green tissue around the main veins (cv. Annette Hegg.).

203 Very pale brown areas develop and coalesce in the yellow-green interveinal areas on the lower leaves (cv. Annette Hegg).
Leaf analysis, 0.10% Mg; 0.73% Mg was found in normal leaves; deficiency, <0.25% Mg; normal range, 0.35–0.75% Mg.

203

204

Iron deficiency

Growth is restricted. The youngest leaves become yellow, and later almost white.

204 The upper leaves become yellow, starting at the base, and growth is restricted; compare with normal plant shown on the left (cv. Annette Hegg).

205 The upper leaves, which are reduced in size, eventually become uniformly pale cream in colour, and small brown areas develop interveinally (centre and right; cv. Annette Hegg).
Leaf analysis, 125 and 89 μg g^{-1} Fe in the normal and deficient leaves respectively.
Determinations of total Fe content do not give a reliable assessment of 'available' Fe in the tissue (cf. p 103.), but values of 100–200 μg g^{-1} Fe are fairly typical for normal growth.

205

Manganese deficiency

Growth is somewhat restricted; the leaves are yellow and become spotted.

206 The leaves become yellow, small brown spots develop interveinally, but the veins remain green (cv. Annette Hegg).
Leaf analysis, 49 and 20 μg g^{-1} Mn in the normal and deficient leaves respectively; deficiency, <25 μg g^{-1} Mn; normal range, 50–200 μg g^{-1} Mn.

206

207

208

209

210

Copper deficiency

The mature leaves become yellow, the margins curl upwards and inwards, and reddish-brown areas develop interveinally.

207 Growth is somewhat restricted, the mature leaves become yellow interveinally and the margins curl upwards and inwards (cv. Annette Hegg).

208 Leaf size is reduced by the deficiency and reddish-brown areas develop interveinally (cv. Annette Hegg). Leaf analysis, 11.6 and 0.4 μg g^{-1} Cu in the normal and deficient leaves respectively; deficiency, <4 μg g^{-1} Cu; normal range, 6–15 μg g^{-1} Cu.

Boron deficiency

The leaves become pale yellow with curled margins. The foliage is brittle and growth is somewhat restricted.

209 The brittle leaves of B deficient plants become pale yellow, except for a green area around the main veins, and the margins curl downwards and inwards. The small veins become reddish-brown and the surrounding tissue eventually dies (cv. Annette Hegg). Leaf analysis, 19 and 31 μg g^{-1} B in the deficient and normal leaves respectively; deficiency <25 μg g^{-1} B; normal range, 30–80 μg g^{-1} B.

Molybdenum deficiency

The plants become stunted, with yellow-green foliage and scorched leaf margins.

210 The yellow-green leaves of Mo deficient plants (right) are greatly reduced in size and the margins become severely scorched (cv. Annette Hegg). Leaf analysis, 2.15 and 0.25 μg g^{-1} Mo in the normal and deficient leaves; deficiency, <0.3 μg g^{-1} Mo; normal range, >0.5 μg g^{-1} Mo (only limited analytical data available).

Index

Plate numbers are shown in bold type

Printed in the UK for HMSO by Linneys Colour Print
Dd 738502 C40 1/87 44942